WHAT, IF ANYTHING, ARE SPECIES?

Species and Systematics

For more information about this series, please visit: https://www.crcpress.com/ Species-and-Systematics/book-series/CRCSPEANDSYS

WHAT, IF ANYTHING, ARE SPECIES?

BRENT D. MISHLER

CRC Press is an imprint of the
Taylor & Francis Group, an **informa** business

First edition published 2021
by CRC Press
6000 Broken Sound Parkway NW, Suite 300, Boca Raton, FL 33487-2742
and by CRC Press
2 Park Square, Milton Park, Abingdon, Oxon, OX14 4RN

Reasonable efforts have been made to publish reliable data and information, but the author and pub-
lisher cannot assume responsibility for the validity of all materials or the consequences of their use.
The authors and publishers have attempted to trace the copyright holders of all material reproduced in
this publication and apologize to copyright holders if permission to publish in this form has not been
obtained. If any copyright material has not been acknowledged please write and let us know so we may
rectify in any future reprint.

Trademark notice: Product or corporate names may be trademarks or registered trademarks and are
used only for identification and explanation without intent to infringe.

ISBN: 9781498714549 (hbk)
ISBN: 9780367715052 (pbk)
ISBN: 9781315119687 (ebk)

Typeset in Times LT Std
by KnowledgeWorks Global Ltd.

Contents

Author

Brent D. Mishler is Director of the University and Jepson Herbaria and Professor in the Department of Integrative Biology at the University of California, Berkeley, where he teaches about island biology, biodiversity, evolution, and phylogenetic analysis. His research interests are in the ecology and evolutionary biology of bryophytes (mosses and liverworts), as well as the theory of phylogenetic systematics. He has been heavily involved in developing electronic resources to present taxonomic and distributional information about plants to the public, with applications to conservation concerns. He has most recently been involved in developing new "spatial phylogenetic" tools for studying biodiversity and endemism using large-scale phylogenies and collection data in a geographic and statistical framework.

1 Introduction

I am amazed how often biologists, who pride themselves on being objective scientists and who criticize the shibboleths of religion, react vehemently when anyone challenges the reality of the rank of species and tries to get rid of binomials. I have heard it so many times:

"We have always had binomials, there would be chaos without them!"

"We must have species in order to do ecology and conservation!"

"How dare you suggest that we get rid of species; they are real entities in the hierarchy of life!"

Species are truly the sacred cow of biology. Most biologists start their consideration of species with an *a priori* assumption that they exist. For example Kunz (2012), in his book-length treatment entitled "Do Species Exist?", never provides any good evidence that they do, he just assumes it. He makes tautological statements like: "If species did not exist, it would not even be possible to speak of the boundaries between them." (p. 12). You could just as easily argue that "if the Tooth Fairy did not exist it would not even be possible to speak of how much she left under a child's pillow." Chung (2004) starts out a paper on the educational value of teaching students controversies over species concepts in biology by flatly taking it as "given that species are real," thus glossing over the most fundamental controversy.

More ink has been spilled on the concept of species than on any other concept in biology. We never seem to eliminate any species concept; rather the field evolves by one concept after another being added to the pile. It is too much to hope that another book on the topic (more ink!) can completely resolve the situation. However, a more modest goal might be achievable, via explaining why there has been such a diversity of views about species and following modern ideas of phylogenetic classification to their logical destination. Perhaps the resolution lies in a different direction than the ever-increasing pile of species concepts. May be we just need to sweep that pile away! May be all the centuries of angst have been due to people striving to define something that does not exist!

SUMMARY

This book will show that some of the persistent furor over species is based on real biology and real differences among organisms. The diversity of views is not semantics, it reflects reality; the units traditionally called species in different organisms *are* different kinds of things and there is no way to make them the same. Just like the rank of genus, the rank of species is applied differently in different cases, in part because of actual biological differences. There is no way to fix this and make the ranks comparable across life. Instead, the modern idea of rankless phylogenetic classification, which is well-established at higher taxonomic levels, needs to be extended to the traditional species level. *Something* is indeed comparable about taxa, including those named as species, but it is not (and cannot be) their ranks.

Biodiversity is not just about species, it instead consists of the entire hierarchy of nested clades representing phylogenetic relationships among all organisms, together with their genetic and functional characteristics, spatial distributions, and ecological relationships. There are many levels of lineages less and more inclusive than the traditional species level. Species and other taxon ranks are not comparable between groups, but lineages and clades are.

ORGANIZATION

The sequence employed in this book follows the author's personal journey in thinking about species over a 40-year period. Nine papers, most co-authored as indicated, are reprinted (with permission), verbatim except that footnotes are renumbered consecutively from the front of the book, figures are renumbered consecutively for each chapter, and a few misspellings are corrected. Newly written material is added to the beginning of each chapter to explain how the reprinted papers in that chapter connect. The discussion is entirely new material, and addresses what the future would look like if my recommendations to get rid of the species rank are followed. Different fonts are used to mark the distinction between newly written text and reprinted text.

I start the book following my initial orientation, which was trying to decide what the species rank should represent under current ranked codes of nomenclature. **Chapter 2** gives a quick review of the history and current variety of species concepts. I argue that some of the apparent chaos among differing ways of viewing species has to do with the fact that the evolutionary processes operating in different branches of the tree of life really are different, and thus specialists in different groups of plants, animals, and fungi have rightly emphasized different criteria when lineages are diverged enough to be called species. My conclusion then, assuming we are going to keep the traditional species rank, was that a pluralistic approach is needed.

In **Chapter 3**, I consider the problem to be how to define species under the current codes of nomenclature. As an early adopter of phylogenetic systematics, my driving principle initially was: if taxa are to be phylogenetic, following the Hennigian revolution of the 1970's and 1980's, then so should species. I and others made attempts in those days to forge a species concept that is compatible with Hennigian phylogenetic systematics or cladistics. Interestingly, nearly every approach to species anyone had ever advocated previously was advocated by one cladist or another. Several such cladistic concepts have been called "the phylogenetic species concept," thus leading to considerable confusion in the literature. Difficulties in arriving at a synthesis include finding the right balance between primary systematic patterns (i.e., character evidence) and evolutionary process theories. Clearly, it makes no sense to apply a species concept that requires prior, specific knowledge of processes (e.g., reproductive behavior or ecological sorting). On the other hand, it is necessary that recognized species taxa be compatible with a phylogenetic system based on descent with modification, if that is to be adopted as the general reference system. One unified *phylogenetic species concept (PSC)* was proposed by me and colleagues in the 1980's, based on a generalized view of the meaning of phylogenetic criteria at any hierarchical level. The grouping criterion was monophyly, but since taxa at all levels are monophyletic, a ranking criterion was needed to decide which monophyletic groups should be named at the rank of species. We felt this ranking criterion had to be pluralistic, with different criteria employed in different biological situations.

The ranking decision needed under this PSC for deciding which phylogenetic groups (clades) should be called species was clearly arbitrary, as was pointed out by friend and foe alike. Continued thinking about this problem prompted me to shift gears. **Chapter 4** covers the next and most radical step I took in my own thinking. Starting in my 1999 paper I recognized that the "species problem" was a special case of the "taxon problem" that was already being addressed for higher taxonomic levels by advocates of phylogenetic nomenclature who wanted to remove ranks from classification. The species rank could be done away with following the same arguments. We need to transition to rankless classification "all the way down," including the rank currently known as species.

In making that transition, however, it is important to ensure that the community can still do everything it is used to doing with species, both practically and theoretically. As discussed above, species have always been seen as fundamental by the general society and by many communities of researchers, such as conservation biologists, ecologists, and population geneticists. Those communities understandably need to be convinced there is a viable alternative before they would ever give up the species rank. Thus **Chapter 5** explores the implications of a rankless phylogenetic approach to terminal taxa in both practice and theory. The ultimate argument is that is it not only *possible* to use rankless classification across biology, but *better*. Rather than imposing artificial conventions like ranks, if our taxonomic practices conform as close as possible to the processes operating in nature to shape biodiversity, we can produce a more useful classification across the board, for everyone from academic biologists to the public.

ACKNOWLEDGMENTS

I thank my co-authors on the papers reprinted in this book, Michael Donoghue (Yale U.), Robert Brandon (Duke U.), Ann Budd (U. Iowa), Edward Theriot (U. Texas), Nico Cellinese (U. Florida), David Baum (U. Wisconsin), and John Wilkins (U. Melbourne) for many intense, fun, and illuminating discussions over the years. They all had a big influence on my thinking, but we don't necessarily agree on everything, and I absolve them from responsibility for the conclusions I draw here. I also thank Kirsten Fisher (Cal State LA) for productive discussions on rankless taxonomy and nomenclature at the level formerly known as species. Chuck Crumly (Taylor and Francis) and Kip Will (UC Berkeley) provided helpful editorial advice and assistance. Open access publication was made possible by support from the Berkeley Research Impact Initiative (BRII) sponsored by the UC Berkeley Library.

LITERATURE CITED

Chung, C. 2004. The species problem and the value of teaching the complexities of species. *The American Biology Teacher* 66: 413–417.
Kunz, W. 2012. *Do Species Exist?* Wiley Blackwell, Weinheim, Germany.

Part I

What Should the Species Level Represent within the Current Ranked Codes of Nomenclature?

2 The Need for Pluralism Because of Different Biologies in Different Taxa

The idea of basic biological "kind" has been with us a long time. There are several good histories (especially Wilkins 2009, 2018), so I will just give a quick summary for my purposes here. Some ancient (but still extant) approaches took a typological approach, relying on logical division with "defining" characters. A slightly more recent view came out of the polythetic "natural system" philosophy of taxonomy developed in the 19th Century (Stevens 1994) which viewed species not as defined by necessary and sufficient characters but as basic clusters of morphological variation – the *"phenetic species concept."* This viewpoint still has many proponents (Levin 1979, Sokal & Crovello 1970, Zapata & Jiménez 2012), surprisingly even among some cladists who view a species as a basic cluster defined by characters with no requirement for evolutionary polarity or monophyly (Nelson & Platnick 1981, Cracraft 1983, Nixon & Wheeler 1990). This view of species as basic clusters of organisms sharing similar traits gained new life with the advent of molecular data: many investigators say they aim to detect "boundaries" between species (e.g., Harrison & Larson 2014, Jain et al. 2018), ironically applying a phenetic species concept with genetic data.

Then there is the equally ancient approach to defining species that has to do with reproductive compatibility going back to folk agricultural observations that like produces like. This approach eventually resulted in the codification of the *"biological species concept"* during the Modern Synthesis. This general approach also had numerous flavors, ranging from the classic isolation approach that emphasized discovery of barriers to interbreeding (e.g., Dobzhansky 1937, Mayr 1996) to newer recognition approach that emphasized whether organisms recognize each other as potential mates or not (e.g., Paterson 1985). Once paleontologists got involved, this concept was extended to add a time dimension, envisioning breeding groups over geological time, the so-called *"evolutionary species concept"* (Wiley 1978). All these views are united in considering interbreeding relationships, or lack thereof, to be the main criteria for defining species.

These two big categories of species concepts both made attempts at "operationality" i.e., providing empirical criteria that a scientist can apply in a practical sense. Those criteria include measuring character variation via morphometrics or clustering DNA samples via RADseq (for applying the phenetic species concept), or doing controlled breeding experiments (for applying the biological species concept).

There have also been purely theoretical species concepts based solely on consid-erations of processes, which did not worry about providing criteria for application. These include such ideas as the *"ecological species concept"* (Van Valen 1976), in which species were viewed as those entities occupying unitary ecological niches. Another theoretical view favored by some philosophers is the concept of *"species as individual"* in which species were viewed as integrated, cohesive units with spatio-temporal boundaries (Ghiselin 1974, Hull 1976). Another theoretical view with no particular operational criteria for application is the *"general lineage concept"* of de Queiroz (1999), which views species as some sort of unitary lineage. In none of these cases was any guidance given as to how one can apply empirical data to decide whether two organisms belong to the same species or not. Some (e.g., Hey 2006) have even argued explicitly that it is a good thing to disregard all the different empirical criteria people have used, in favor of a purely theoretical "unified" view of species (begging the question about the utility of a theoretical construct with no application to the real world). Furthermore, in none of these author's arguments was there a clear distinction between the species level and groupings at other levels, i.e., there could be groups filling ecological niches, or making up lineages, that are nested at more than one hierarchical level.

Finally, there have been various "phylogenetic" species concepts proposed. They are a heterogeneous lot – as noted above, some of these (e.g., Cracraft 1983, Nixon & Wheeler 1990) are really phenetic in that species are regarded as a cluster of organ-isms that is homogeneous for characters and monophyly is explicitly ruled out by definition. That sort of species concept is clearly not phylogenetic and thus seems both misnamed and an unlikely basis for a phylogenetic system of classification. Another *"phylogenetic species concept,"* the one I will refer to by that name in this book, is the one I and co-authors developed, which tries to give both theoretical and operational criteria for defining species phylogenetically. In this view, species are the basal-most monophyletic groups that are named taxonomically. That concept will be discussed in detail in the following chapter.

The goal of *this* chapter is to address the question of why so many different con-cepts have been proposed. The first paper reprinted below (Mishler & Donoghue 1982) argued that the primary reason for the existence of a species problem is that the species concepts and criteria outlined above conflict in most real cases – different concepts (and processes) "pick out" different groups in each particular case. In other words, the interbreeding groups often do not match the phenetic clusters, or the sets of organisms filling the same niche. If all species concepts led to recognition of the same entities, then there would not have been much controversy. But as it turns out, the differences among biologists promoting different species concepts are not pure semantics, bias, or stubbornness – it reflects a fundamental biological truth.

The processes influencing divergence of lineages are manifold, vary considerably in their action from group to group, and are often acting at cross purposes to each other. With the application of rapidly improving molecular tools, the recent literature provides an increasing number of examples of this heterogeneity in evolutionary processes, even in vertebrates, as well as in plants and microbes where it has been known for awhile. *Cessation of interbreeding turns out often to not be the most important factor in primary divergence of lineages*, despite what Mayr and others argued. Thus the implied correspondence between interbreeding groups and groups defined by other criteria, relied on by many species concepts and explicitly stated by De Queiroz (1999), has been abundantly falsified.

The second paper reprinted below (Mishler 1985) brought in developmental constraints as another class of causal processes influencing the cohesion and divergence of lineages, that up until that point had not been introduced to the species debate, even though it was being hotly debated as an explanation for higher-level patterns of lineage divergence in macroevolution (Alberch 1980). Evolutionary divergence in phenotypes often follows lines of least resistance in modifications of developmental programs, and this could often be true at the primary divergence level as well.

The third paper reprinted below (Mishler & Budd 1990) is an introduction to a whole symposium (*Systematic Botany* 15:1) that addressed natural experiments that have played out over and over in both plants and animals. Asexual groups of organisms (dismissed by Mayr and others as irrelevant aberrations to the biological species concept) actually provide excellent study systems for looking at the effects of patterns of interbreeding on divergence. If there are distinctly differentiated lineages in asexual groups, then it must be due to processes other than interbreeding or cessation thereof, such as ecological or developmental constraints. Whether asexual reproduction is an evolutionary dead-end in the long run, or not, is a completely different issue. While it lasts, asexuality provides an invaluable way to factor out one of the big contenders in the species debates and detect how important the other processes might be. Asexual lineages are deserving of much more study, in comparison to their sexual relatives, in this regard.

LITERATURE CITED

Alberch, P. 1980. Ontogenesis and morphological diversification. *American Zoologist* 20: 653–667.

Cracraft, J. 1983. Species concepts and speciation analysis. *Current Ornithology* 1: 159–187.

Dobzhansky, T. 1937. What is a species? *Scientia* 61: 280–286.

de Queiroz, K. 1999. The general lineage concept of species and the defining properties of the species category. Pp. 49–88 in: *Species, New Interdisciplinary Essays*, R. A. Wilson (ed.). Bradford/MIT Press.

Ghiselin, M.T. 1974. A radical solution to the species problem. *Systematic Zoology* 23: 536–544.

Harrison, R.G. and E.L. Larson. 2014. Hybridization, introgression, and the nature of species boundaries. *Journal of Heredity* 105: 795–809.

Hey, J. 2006. On the failure of modern species concepts. *Trends in Ecology & Evolution* 21: 447–450.

Jain, C., Rodriguez, L.M., Phillippy, A.M. Adam, K.T. Konstantinidis, and S. Aluru. 2018. High throughput ANI analysis of 90K prokaryotic genomes reveals clear species boundaries. *Nature Communications* 9: 5114.

Levin, D.A. 1979. The nature of plant species. *Science* 204: 381–384.

Mayr, E. 1996. What is a species, and what is not? *Philosophy of Science* 2: 262–277.

Mishler, B.D. and M.J. Donoghue. 1982. Species concepts: a case for pluralism. *Systematic Zoology* 31: 491–503.

Mishler, B.D. 1985. The morphological, developmental, and phylogenetic basis of species concepts in bryophytes. *The Bryologist* 88: 207–214.

Mishler, B.D. and A.F. Budd. 1990. Species and evolution in clonal organisms–introduction. *Systematic Botany* 15: 79–85.

Nelson, G.J. and N.I. Platnick. 1981. *Systematics and Biogeography: Cladistics and Vicariance*. Columbia University Press, New York.

Nixon, K.C. and Q.D. Wheeler. 1990. An amplification of the phylogenetic species concept. *Cladistics* 6: 211–223.

Paterson, H.E.H. 1985. The recognition concept of species. Pp. 21–29 in: *Species and Speciation*, ed. E. S. Vrba. Transvaal Museum, Pretoria, South Africa.

Sokal, Robert R. and T. Crovello. 1970. The biological species concept: A critical evaluation. *American Naturalist* 104: 127–153.

Stevens, P.F. 1994. *The Development of Biological Systematics*. Columbia University Press, New York.

Van Valen, L. 1976. Ecological species, multispecies, and oaks. *Taxon* 25: 233–239.

Wiley, E.O. 1978. The evolutionary species concept reconsidered. *Systematic Zoology* 27: 17–26.

Wilkins, J.S. 2009. *Species: A History of the Idea*. University of California Press, Berkeley.

Wilkins, J.S. 2018. *Species: The Evolution of the Idea*. CRC Press, Boca Raton.

Zapata, F. and I. Jiménez. 2012. Species delimitation: inferring gaps in morphology across geography. *Systematic Biology* 61: 179–194.

Species Concepts: A Case for Pluralism[1]

"We must resist at all costs the tendency to superimpose a false simplicity on the exterior of science to hide incompletely formulated theoretical foundations."

(Hull, 1970:37)

It has often been argued that it is empirically true and/or theoretically necessary that "species," as units in nature, are fundamentally and universally different from taxa at all other levels. Species are supposed to be unique because they are individuals (in the philosophical sense, as opposed to classes) – integrated, cohesive units, with a real existence in space and time (Ghiselin, 1974; Hull, 1978). Interbreeding among the members (parts) of a species and reproductive isolation between species are generally believed to account for their individuality. These reproductive criteria are supposed to provide the greater objectivity of the species category and have been suggested as *the* criteria by which species taxa are to be delimited in nature.

Wake (1980) has pointed out that this conception of species forms the basis upon which Eldredge and Cracraft (1980) have built their formulation of evolutionary process and phylogenetic analysis. In fact, this notion of species seems to underlie much of the recent and growing body of theory which, for convenience, could be called macroevolutionary theory (Eldredge and Gould, 1972; Stanley, 1975; Gould, 1982). Moreover, most recent texts in systematics and ecology are predicated on the idea that species taxa are unique and fundamental (e.g., White, 1978; Ricklefs, 1979; Wiley, 1981). It is therefore important to assess carefully any claim that species do or should possess the properties of individuals, and whether breeding criteria are adequate indicators of individuality.

The "species problem" has yielded an enormous quantity of literature, and it is not the purpose of this paper to provide a review (for which see Mayr, 1957; Wiley, 1978; and papers cited therein). Instead, we will (1) briefly characterize prevailing species concepts, (2) summarize some empirical observations that bear on the species problem, (3) consider the respects in which species taxa as currently delimited by systematists do and do not have the properties of individuals, (4) discuss several choices with which we are faced if all the criteria of individuality are not always met.

We will argue that current species concepts are theoretically oversimplified. Empirical studies show that patterns of discontinuity in ecological, morphological, and genetical variation are generally more complex than are represented by

[1] B.D. Mishler and M.J. Donoghue. 1982. Species concepts: a case for pluralism. Systematic Zoology 31: 491–503. [reprinted by permission]

these concepts. Criteria for what constitutes "important" discontinuity appear to vary in response to the vast differences in biology between groups of organisms. In our view, no single and universal level of fundamental evolutionary units exists; in most cases, species taxa have no *special* reality in nature. We urge explicit recognition and acceptance of a more pluralistic conception of species, one that recognizes the evident variety and complexity of "species situations." We will conclude by exploring important consequences of this view for ecology, paleontology, and systematics.

Prevailing Species Concepts

A consensus appears to have been reached that species are integrated, unique entities. The so-called biological species concept emphasizes that species are reproductive communities within which genes are (or can be) freely exchanged, but between which gene flow does not occur or at least is very rare (e.g., Mayr, 1970). According to this view, a species is a group of organisms with a common gene pool that is reproductively isolated from other such groups.

The evolutionary species concept (Simpson, 1961; Grant, 1971; Wiley, 1978, 1981) is an important extension of the concept of biological species, an attempt to broaden the definition to include all sorts of organisms (not just sexually reproductive ones) and to portray the existence of species through time. According to this view, species are separate ancestor-descendant lineages with their own evolutionary roles, tendencies, and fates. The ecological species concept of Van Valen (1976) is similar (but see Wiley, 1981), however, it emphasizes the "adaptive zone" occupied by a lineage.

Ghiselin (1974) and Hull (1976, 1978) have examined the status of species from a philosophical standpoint. They contend that if species are to play the role required of them in current systematic and evolutionary theory, they must be "individuals" (i.e., integrated and cohesive entities with a restricted spatio-temporal location) rather than "classes" (i.e., spatiotemporally unrestricted sets with defining characteristics). Hull (1980), Wiley (1980, 1981), and Ghiselin (1981) argue that species are fundamentally different from genera, families, and other higher taxa because they are the most inclusive entities that are "actively evolving."

In general, then, species are considered to be the most objectively defined taxonomic and evolutionary units. As Mayr (1970:374) put it, they are "the real units of evolution, as the temporary incarnation of harmonious, well-integrated gene complexes." They differ from taxa at all other levels, which are considered to be arbitrarily defined and more subjective categories (e.g., Mayr, 1969:91).

For many workers, these views are not only theoretically satisfying but also seem sufficiently unproblematical in application. Many biologists (especially zoologists) seem to be satisfied that, with the exception of some sibling species complexes and rassenkreise, the application of biological/evolutionary species concepts will yield the same sets of organisms that would be recognized as "species" by a competent taxonomist in a museum, or by a person on the street.[2]

It must be pointed out, however, that the prevailing species concepts are based on relatively few well-studied groups such as birds and *Drosophila*, groups in which discontinuities in the ability to interbreed are relatively complete, and discontinuities in morphological and ecological variation coincide well with the inability to breed in nature. It also must be pointed out that even though relatively few groups have been studied in detail, a correspondence between morphological, ecological, and breeding discontinuities is often simply assumed.

The acceptance of biological/evolutionary species concepts has not been universal. In particular, the botanical community has not wholeheartedly taken them up, and alternatives have been proliferated.[3] It seems clear that the group

---

[2] Gould (1979) and others have defended the biological species concept on the grounds that the same taxa recognized by western taxonomists are recognized by tribespeople in New Guinea, etc. There are several problems with this kind of argument. First, it is not clear that this finding constitutes an independent test because, after all, New Guinea tribespeople are human too, with similar cognitive principles and limitations of language. It should also be borne in mind that the observer is by no means neutral. Folk taxonomies have been collected by people with a knowledge of evolution and modern systematic concepts. Second, it is generally not a strong argument to show that a pre-scientific society has recognized something that modern science currently accepts. Surely a modern astronomer would not consider it very strong evidence that a primitive mythology supported one cosmological theory over another. Finally, the taxa recognized by western taxonomists (and often by natives at some level of their linguistic hierarchy) in these instances are not known to be biological species – for the most part they are morphological units that are believed to be reproductively isolated from other such units.

[3] Initially, the biological species concept was embraced and promulgated by plant systematists interested in evolution (Stebbins, 1950; Grant, 1957). Cronquist (1978) detailed Grant's efforts (from 1956 to 1966) to apply the biological species concept in *Gilia* (Polemoniaceae). It very soon became apparent that the biological species concept was fraught with difficulties, but Grant chose to amend the concept (rather than abandon it altogether), first (1957) with the notion of the syngameon (i.e., the unit of interbreeding higher than the species), later (1971) by adopting an evolutionary species concept. Finally, in the second edition of his classic book on plant speciation, Grant (1981) treats species in a more flexible and pluralistic manner. Some botanists (e.g., Stebbins, 1979:25) continue to feel that the biological species concept, or some modification of it, is the only suitable framework for understanding plant diversity. However, many (perhaps most) botanical systematists remain rather skeptical about the general applicability of the concept in botany (Davis and Heywood, 1963; Raven, 1976; Cronquist, 1978; Levin, 1979; Stevens, 1980a).

The different attitudes of zoologists and botanists towards the concept of species may be of interest to historians, sociologists, and philosophers of science. For organismic and evolutionary biology, the "modern synthesis" of the 1930's and 1940's may have represented a revolution in the sense of Kuhn (1970). For systematists, the principal outcome was the biological species concept. Zoologists (especially vertebrate systematists) appear to have largely accepted the new paradigm and to have entered a period of "normal science/' applying the concept in particular cases ("puzzle-solving"). While problems like sibling species, semispecies, and subspecies have become apparent, these have generally not prompted a critical evaluation of the paradigm or a proliferation of alternatives. In contrast, in the botanical community the biological species concept was soon found to be inapplicable or of difficult application and likely to lead to confusion. This resulted in a groping for alternatives and a defense of older concepts. In this regard, the historical development of species concepts in botany seems to fit better. Feyerabend's (1970) characterization of scientific change as the simultaneous practice of normal science and the proliferation of alternative theories.

of organisms on which one specializes strongly influences the view of "species" that one develops. It also seems clear that in order to fully appreciate biological diversity (for purposes of developing general concepts), it is essential to study a variety of different kinds of organisms, or at least take seriously those who have.[4]

Numerous attacks have been leveled at the biological/evolutionary species concepts. Many of these have been concerned primarily with whether they are operational (e.g., Sokal and Crovello, 1970). However, as Hull (1968, 1970) has pointed out, a concept cannot be completely operational and still be useful for the growth of science. The critical question is whether a concept is operational *enough* to be useful as a conceptual framework. Considerations of operationally, while certainly of interest, are not central to the argument developed below, which primarily concerns the theoretical adequacy of prevailing species concepts.

Empirical Considerations

In our view, a theoretically satisfactory species concept must bear some specifiable relationship to observed patterns of variation among organisms. It is *not* acceptable to adopt a definition of species simply because it conveniently fits into some more inclusive theory, e.g., a theory of evolutionary process. A species concept is, in effect, a low-level hypothesis about the nature of that variation, itself subject to empirical tests. Therefore, in this section, we summarize some relevant empirical findings, many of which have not been generally recognized.

The Noncorrespondence of Discontinuities

The reason for discontent among botanists and other workers is not that they have been unable to perceive discontinuities in nature. Instead, it has become apparent that there are many kinds of discontinuities, all of which may be of interest (Davis and Heywood, 1963:91). The question is, how well do various discontinuities correspond; i.e., are the same sets of organisms delimited by discontinuities when we look at morphology, as when we look at ecology, or breeding? The answer appears to be that there is no necessary correspondence. Stebbins (1950), Grant (1957, 1971, 1981), Stace (1978), and many others have discussed hybridization, apomixis, polyploidy, and anomalous breeding systems in plants and have clearly documented the frequent noncorrespondence of different kinds of discontinuities. In some groups, there is complete reproductive

---

[4] The zoologists initially responsible for developing the biological species concept were aware of the difficulties in applying the concept in some groups of animals and many groups of plants. Dobzhansky (1937, 1972) consistently pointed out the diversity of "species situations" observable in nature. Mayr (1942:122) was careful to point out differences between plants and animals, and difficulties in the practical application of the biological species concept in some cases. Particularly, rigid versions of the biological species concept have been promulgated more recently, in attempted generalizations that have shown a startling lack of concern for the biology of the majority of organisms on earth. Mayr (1982) has examined the resistance of botanists to the biological species concept and concluded that "the concept does not describe an exceptional situation" (p. 280). But he grants some justification to the ideas of "certain botanists" who question "whether the wide spectrum of breeding systems that can be found in plants can all be subsumed under the single concept (and term) 'species'" (p. 278).

isolation between populations that would be recognized as one species on morphological grounds (i.e., "sibling species," as in some groups of *Gilia* (Grant, 1964), and *Clarkia* (Small, 1971)), and in many other groups of plants the interbreeding unit encompasses two to many morphological units (e.g., *Quercus* (Burger, 1975)).

It has also become clear that discontinuities in morphological variation or in the ability to interbreed do not necessarily correspond to differences in ecology ("niche?"). The early work of Turesson (1922a, 1922b) in Europe, and of Clausen, Keck, and Hiesey (1939, 1940) in North America, demonstrated that ecotypes "may or may not possess well-marked morphological differences which enable them to be recognized in the field" (Stebbins, 1950:49). The great extent to which local populations of the same biological or morphological species are physiologically differentiated and adapted to their particular environments is only now being realized (Mooney and Billings, 1961; Antonovics et al., 1971; Antonovics, 1972; Bradshaw, 1972; Kiang, 1982).

If noncorrespondence is prevalent, then strict biological species will not necessarily have anything in common but reproductive isolation. It might be argued that a species concept that unambiguously reflects one aspect of variation may be preferable to one that ambiguously reflects several things. But why should we necessarily pin species names on sets of organisms delimited by reproductive barriers? Why not choose, for example, to name morphological units instead?

One argument for pinning species names on reproductively isolated groups is that breeding discontinuities are thought to be more clear-cut than morphological ones and therefore less arbitrary. However, Ornduff (1969) has summarized the complexity of the reproductive biology of flowering plants and pointed out the difficulty of applying rigid species delimitations based on interfertility. When variation in the ability to interbreed is examined in detail, we find discontinuities of many different degrees and kinds. Groups of organisms range from completely interfertile to completely reproductively isolated. Hierarchies or networks of breeding groups vary in complex ways in space and time. Therefore, even if we were to decide that breeding discontinuities were theoretically the most important kind of discontinuity, and the ones that species names should reflect, the choice of what constitutes a significant discontinuity remains problematical.

A second argument for the importance of reproductive barriers is that gene flow prevents significant divergence while a lack of gene flow allows it. However, this now appears not to be the case. If a population is subjected to disruptive selection, there can be divergence even in the face of gene flow (Jain and Bradshaw, 1966). In these instances, it appears that some means of reproductive isolation will usually evolve, but such isolation follows initial divergence. Moreover, allopatric populations can remain morphologically similar for very long periods or they can diverge morphologically (see discussion of this point by Bremer and Wanntorp, 1979a). This morphological divergence may or may not be accompanied by reproductive isolation, though it appears likely that eventually, a reproductive barrier will result. The point is that morphological divergence and the attainment of means of reproductive isolation can be uncoupled events in time and space. Levin (1978:288–289) concluded:

> "If we adhere to the biological species concept—the integrated reproductive com-munities—described by Mayr, then speciation is capricious ... Isolating mechanisms are not the cause of divergent evolution, nor are they essential for it to occur."

A related, larger-scale argument for the importance of reproductive barriers is that groups that are reproductively isolated for long periods of time are at least evolutionarily independent (whether or not they diverge morphologically), mak-ing them effectively separate entities. Reproductive barriers indeed may often be important in this way, but other factors such as ecological role and homeostatic "inertia" are important as well. Because of the complex nature of variation in each of these factors, and because different factors may be "most important" in the evolution of different groups, a *universal* criterion for delimiting fundamen-tal, cohesive evolutionary entities does not exist.

Questionable Internal Genetic Cohesion

The notion of integration and internal cohesion is central to biological/evolution-ary/individualistic species concepts. In this paper, we will follow the common assumption that "cohesion" means genetic cohesion maintained via gene flow, a notion that has recently been explicitly formulated (Wiley and Brooks, 1982). However, Hull (1978) has pointed out that other factors such as internal homeo-stasis and "external environment in the form of unitary selection pressures" (p. 344) may contribute to or confer cohesion. It seems to us likely that "cohesion," and the factors responsible for it, will differ from one group of organisms to another and from one level in the hierarchy to another.

Ehrlich and Raven (1969) pointed out that the extent of gene flow seems to be very limited in many organisms and may not account for the apparent integrity of the morphological units we recognize in nature. Bradshaw (1972:42) suggested that "effective population size in plants is to be measured in meters and not in kilometers." Endler (1973) studied clinal variation and concluded that "gene flow may be unimportant in the differentiation of populations along envi-ronmental gradients" (p. 249). Levin and Kerster (1974) thoroughly reviewed and analyzed the literature concerning gene flow in seed plants and concluded that "the numbers [of individuals] within panmictic units are to be measured in tens and not hundreds" (p. 203). These same points were reiterated by Sokal (1973), Raven (1976), and Levin (1978,1981). Levin (1979:383) stated:

> "The idea that plant species are Mendelian populations wedded by the bonds of mating is most difficult to justify given our knowledge about gene flow. Indeed a contrary viewpoint is supported. Populations separated by several kilometers may rarely, if ever, exchange genes and as such may evolve independently in the absence of strong or even weak selective differentials."

Lande (1980) has stressed that there has been an overemphasis on the genetic cohesion of widespread species and argued that "of the major forces conserving phenotypic uniformity in time and space, stabilizing selection is by far the most powerful" (p. 467). Grant (1980:167) suggested that "the homogeneity of species is due more to descent from a common ancestor than to gene exchange across significant parts of the species area."

Jackson and Pound (1979) critically reviewed much of this literature and rightly pointed out that there is little rigorous evidence in animals to support or to reject the generality of any statement about gene flow because detailed studies are rare. They concluded, however, that data "seem sufficient to indicate that gene flow in plants can be limited due to local or leptokurtic dispersal of pollen and seeds" (p. 78). It is important to keep in mind that population genetic theory predicts that a small amount of migration between populations may be sufficient to maintain genetic similarity in the absence of differential selection (Lewontin, 1974:212–216). Clearly, determining the relative importance of factors such as gene flow, developmental homeostasis, and selection in nature will require rigorous population genetic theory (e.g., Lande, 1980) and careful quantification of empirical data, rather than qualitative, anecdotal arguments.

Evolutionary biologists are just beginning to understand gene flow in plants and animals, but have hardly begun to address the complicated patterns of gene exchange present in the fungi, bacteria, and "protists." A kind of chauvinism has so far restricted discussions of gene flow to comparisons of biparental sexual organisms and asexual ones. Complex patterns of sexuality are present in the fungi (Clémençon, 1977); intricate incompatibility systems, as well as incompletely understood parasexuality cycles, make the simplistic application of the biological species concept impossible in most cases. The existence of discrete, integrated genetic lineages is even less likely in the "Monera" (Cowan, 1962). There probably are very few absolute barriers to genetic exchange in bacteria, because of the phenomena of DNA-mediated transformation, phage-mediated transduction, and bacterial conjugation (Bodmer, 1970).

Are Species Taxa Individuals?

In our view, the empirical considerations discussed above indicate that in many (perhaps most) major groups of organisms, actual patterns of variation are such that the species taxa *currently recognized* by taxonomists cannot be considered discrete, primary, and comparable "individuals," integrated and cohesive via the exchange of genes, fundamentally different from taxa at other levels. Variation in morphology, ecology, and breeding is enormous and complex; there are discontinuities of varying degree in each of these factors and the discontinuities are often not congruent. There may often be roughly continuous reduction in the degree of cohesion due to gene flow as more inclusive groups of organisms are considered. The acquisition of reproductive isolating mechanisms appears in many cases to be fortuitous and such isolation is neither the cause of morphological or ecological divergence nor is it necessary for divergence to occur.

Although many currently recognized species do not meet one important criterion of "individuality," namely cohesion and integration of parts, another important criterion often is met, namely restricted spatiotemporal location (i.e., units united by common descent). These units are not strictly "individuals" or "classes," but clearly they can function in evolutionary theory and phylogeny construction. Wiley (1980) called such units "historical entities," but applied this term only to taxa above the species level.

We should mention, as a disclaimer, that although many species taxa (as currently delimited) cannot be considered unique, individualistic units, this does not mean that all species taxa are not. In some groups of organisms, biological species may conform in all respects to the philosophical concept of individual. We simply suggest that this condition is a "special case," and that unwarranted extrapolations have been made from a very few groups of organisms to organisms generally.

Some Options

As discussed above, in many plant and some animal groups, evolutionary processes (i.e., replication and interaction in the sense of Hull, 1980) occur primarily on a small scale (even when extrapolated over many generations) relative to the traditional species level. In such groups, the units in nature that are more like individuals are actually interbreeding local populations, and therefore, the basal taxonomic unit (the species) is currently more inclusive than the basal evolutionary units (the populations). This means that many presently recognized species taxa are, at best, historical entities. If this is the case, and if we want species taxa that are more fully individuals, can we bring taxonomic practices in line with our theoretical desires, and at what cost? If we cannot, or if the costs are too great, are there any theoretically acceptable alternatives, and what would they entail?

We formulate here three options with which we are faced and reject the first two. In the next section, we explore some implications of the third alternative.

1. Alter the usage of "species" to equal "evolutionary unit," i.e., attempt to locate all of the effectively isolated and independently evolving populations and apply species names to them.
2. Alter the usage of "species" to equal the "cenospecies" or "comparium" (see Stebbins, 1950; Grant, 1971), i.e., recognize as the basic taxonomic units only those taxa that are *completely* intersterile.
3. Apply species names at about the same level as we have in the past, and decouple the basal taxonomic unit from notions of "basic" evolutionary units.

We reject choice (1) for several reasons, some practical and some theoretical. In a practical sense, formally naming whatever the truly genetically integrated units turned out to be would be disastrous. There are certainly very many such units, they are at best very difficult to perceive even with the most sophisticated techniques and in the most studied organisms, and these units are continuously changing in size and membership from one generation to the next. At any one time, we can never know which units will diverge forever.

Rosen (1978, 1979) has discussed and adopted a species concept quite similar to choice (1). While we would generally agree with him that populations with apomorphous character states are units of evolutionary significance (1978:176), we could not agree that species should be "the smallest natural aggregation of individuals with a specifiable geographic integrity that can be defined by any current set of analytical techniques" (1979:277). Since we could probably distinguish each individual organism, or very small groups of organisms, on the basis of apomorphies (if we looked hard enough), why shouldn't each of these units be given a Linnaean binomial?

There is a more important, theoretical reason for rejecting alternative (1), one that we have alluded to above. A pervasive confusion runs through much discussion of species: the erroneous notion that *a* single basal evolutionary unit is somewhere to be found among all the possible units that could be recognized. There are *many* evolutionary, genealogical units within a given lineage (Hull, 1980) – a rough hierarchy or network of units, which may be temporally and spatially overlapping. Thus, in the search to find *the* evolutionary unit, one is on a very "slippery slope" indeed. Units all along this slope may be of interest to evolutionists, depending on the level of focus of the particular investigator. These units do require some sort of designation in order to be studied, but a formal, hierarchical Linnaean name is not necessary.

Option (2), in many instances, would represent the opposite extreme (an attempt to locate the "top" of the slippery slope). Absolute reproductive isolation would be used as the overriding ranking criterion. If two organisms could potentially exchange genes, either directly or through intermediates, they would be placed in the same species taxon. There are several reasons why we reject this alternative.

First, it is unclear that reproductive criteria necessarily provide species taxa that are useful for purposes of phylogeny reconstruction and historical biogeography. As Rosen (1978, 1979) and Bremer and Wanntorp (1979a) have pointed out, "biological species" may be paraphyletic assemblages of populations united only by a plesiomorphy, i.e., all those organisms that have not acquired a means of reproductive isolation. If reproductive criteria are to be useful for cladistic analysis, it is necessary to determine which modes of isolation arose as evolutionary novelties in a group.

Our second objection to option (2) has to do with the problem of measuring "potentiality." There have been numerous comments on the inadequacy of potential interbreeding as a ranking criterion, and even strong proponents of the biological species concept have rejected potential interbreeding as a part of their species definitions. Under certain conditions, very disparate organisms can be made to cross. If we adopted this option, the family Orchidaceae, with approximately 20,000 species at present (covering a great range of variation), might be lumped into just a few species because horticulturalists have produced so many bi-and pluri-generic hybrids. The universal application of any one criterion will undoubtedly obscure important patterns of variation in other parameters.

Species Like Genera

If we adopted alternative (3), what would happen to the species category? Would species taxa necessarily be theoretically meaningless entities? Are all alternatives to biological/evolutionary/individual species concepts devoid of theoretical interest as implied by Eldredge and Cracraft (1980:94)?

We would agree that if species were simply phenetically similar groups of populations they might indeed be unsatisfactory for many purposes. The application of species concepts like those of Cronquist (1978) and Nelson and Platnick (1981) may yield species taxa that are not useful from the standpoint of reconstructing phylogenies (see discussion by Beatty, 1982).[5]

However, we think that one form of option (3) may provide theoretically meaningful units. In groups where the actual interbreeding units are small relative to the morphologically delimited units, species can be considered to be like genera or families or higher taxa at all levels. That is, they are assemblages of populations united by descent just as genera are assemblages of species united by descent, etc. If we required that species be monophyletic assemblages of populations (to the extent that this could be hypothesized), then they could play a role in evolutionary and phylogenetic theory just as monophyletic taxa at all levels can. Theoretical significance does not reside solely in the basal taxonomic units or in units that are "fully individuals."

If we recognize that species are like genera, and insist that they be monophyletic, then we are faced with the problems of assessing monophyly and of ranking, problems that plague systematists working at all levels. Several different concepts of monophyly have been employed by systematists, but none of them explicitly at the species level (see discussion by Holmes, 1980). We favor Hennig's (1966) concept of monophyly (except explicitly applied at the species level) but are fully aware of the difficulties in its application at low taxonomic levels (Arnold, 1981; Hill and Crane, 1982). In particular, the difficulty posed by reticulation (hybridization) (Bremer and Wanntorp, 1979b) may be especially acute at lower taxonomic levels. Using synapomorphy as evidence of monophyly requires that the polarity of character states be determined, and again this may be an especially difficult problem near the species level. Polarity assessments will be possible to a greater or lesser extent depending on the certainty with which out-groups are known (Stevens, 1980b).

As noted previously, in order to use reproductive isolation as evidence of monophyly, it would be necessary to determine which means of reproductive isolation are apomorphies at a given level, and which are not. An example of the difficulty of applying a Hennigian concept of monophyly is the very real possibility of "paraphyletic speciation." If speciation by peripheral isolation happens frequently, then a population (geographically defined), which has developed some apomorphic feature (such as a morphological novelty or an isolating mechanism) with respect to its "parent" species, may often be cladistically more closely related to some part of the parent species than to the remainder (see

---

[5] The species concepts of Cronquist and of Nelson and Platnick are as follows:

Cronquist (1978:15): "the smallest groups that are consistently and persistently distinct, and distinguishable by ordinary means."

Nelson and Platnick (1981:12): "the smallest detected samples of self-perpetuating organisms that have unique sets of characters."

discussion and example in Bremer and Wanntorp, 1979a). In such a case, we would take the (perhaps controversial) position that if the population is to be recognized as a formal species taxon, and if the phylogenetic relationships of the populations in the parent species can be resolved, then the taxonomist should not formally name the parent "species" (which has now been found to be a paraphyletic group), but instead name monophyletic groups discerned within it. Conversely, however, if cladistic structure within the parent species cannot be resolved, then in our view it would be acceptable to provisionally name it as a species (even if the populations included within shared no apomorphy).

This example illustrates the fact that even when monophyletic groups are delimited, the problem of ranking remains since monophyletic groups can be found at many levels within a clade. Species ranking criteria could include group size, gap size, geological age, ecological or geographical criteria, degree of intersterility, tradition, and possibly others. The general problem of ranking is presently unresolved, and we suspect that an absolute and universally applicable criterion may never be found and that, instead, "answers" will have to be developed on a group by group basis.

Some Consequences of Pluralism

We have outlined a concept of species (i.e., "species like genera") that may be appropriate for groups of organisms in which certain conditions obtain. However, we think that a variety of species concepts are necessary to adequately capture the complexity of variation patterns in nature. To subsume this variation under the rubric of any one concept leads to confusion and tends to obscure important evolutionary questions. As Hull (1970; see epigraph) has argued, we must resist the urge to superimpose false simplicity. If "species situations" are diverse, then a variety of concepts may be necessary and desirable to reflect this complexity.

Many theories in biology appear to lack the universality of theories in other natural sciences. Often the problem is to decide which one of several theories (not necessarily mutually exclusive) applies to a particular situation (for a specific application of this theoretical pluralism to evolutionary biology, see Gould and Lewontin, 1979). A satisfactory general theory is one in which the number of sub-theories is kept to a minimum, but not reduced to the point where important patterns and processes are obscured. The evaluation of how well a theoretical system "accounts for" patterns in the world is problematical, and we cannot offer any generally applicable criterion for making such an evaluation. However, in the case of species, we think that the search for a universal species concept, wherein the basal unit in evolutionary biology and in taxonomy is the same, is misguided. In our opinion, it is time for "species" to suffer a fate similar to that of the classical concept of "gene."[6]

---

[6] Initially, the "gene" was considered to be the unit of heredity, but the classical concept of gene has been replaced by several concepts which stand in a complex relation to one another (Hull, 1965). The use of a disjunctive definition (Hull, 1965) allows a single term to designate a complex of concepts. However, this can become so confusing that it may be desirable to replace (at least in part) an old terminology with a new set of terms with more precise meanings.

We should recognize that species taxa have never been, and very probably cannot be made readily comparable units. This observation has a number of important theoretical implications. Ecologists must consider the extent to which "species" can be considered equivalent and comparable from one group of organisms to another. Population sizes and structures, gene flow, social organization, the nature of selective factors, and developmental constraints differ in multifarious ways. This means that it is imperative that systematists be explicit about the nature of variation in, and the properties of, the species that they recognize in the groups they study. In turn, the users of species names must at all times be aware that "species are only equivalent by designation, and not by virtue of the nature or extent of their evolutionary differentiation" (Davis and Heywood, 1963:92). As obfuscatory as this may seem, comparative biologists must not make inferences from a species name without consulting the systematic literature to see what patterns of variation the name purports to represent.

These considerations are also important to paleontologists, who make inferences about, and from, "fossil species," and imply correspondences between variation in morphology, ecology, and breeding. It is perplexing that some quite innovative paleontologists, such as Eldredge and Gould, have uncritically retained the biological species concept in their work. As we have shown, there are many reasons why species should not be treated as particles or quanta. Paleontologists should consider exactly what macroevolutionary theories require species to be. For many purposes, they may not require species that are completely individuals, but simply monophyletic lineages. If units that are cohesive via gene flow are an absolute requirement, then fossils may not provide appropriate evidence.

Finally, what are the implications for the systematist of a pluralistic outlook on species? Systematists working on relatively little known organisms should not assume that concepts derived from other groups of organisms are necessarily applicable. Instead, in each group, the systematist is obligated to study patterns of variation in morphology, ecology, and breeding, and to detail the nature of the correspondences among these patterns. It is essential that the ways in which names are applied to taxa at all levels be stated explicitly.

If we adopt a case by case approach and urge specialists to unabashedly develop concepts for their particular groups, are we saying that "anything goes?" Of course, the answer is no. We are only suggesting pluralism within limits. Taxa (including species) recognized by systematists must have a specifiable relationship to theoretically important variation, more specifically, we have argued that species taxa should be phylogenetically meaningful units. There may not be a *universal* criterion to arbitrate between conflicting species classifications of a given genus, but through the complex process that is science, the community of involved workers can and will hammer out criteria for making such decisions.

ACKNOWLEDGMENTS

We are indebted to J. Beatty, E. Coombs, S. Fink, W. Fink, C. Hill, D. Hull, E. Mayr, N. Miller, P. Stevens, and five anonymous reviewers for criticizing this manuscript at one stage or another during its ontogeny; however, they are not to blame for its contents.

REFERENCES

ANTONOVICS, J. 1972. Population dynamics of the grass *Anthoxanthum oderatum* on a zinc mine. J. Ecol., 60:351–365.

ANTONOVICS, J., A. D. BRADSHAW, AND R. G. TURNER. 1971. Heavy metal tolerance in plants. Adv. Ecol. Res., 7:1–85.

ARNOLD, E. N. 1981. Estimating phylogenies at low taxonomic levels. Z. Zool. Syst. Evolut.-Forsch., 19:1–35.

BEATTY, J. 1982. Classes and cladists. Syst. Zool., 31:25–34.

BODMER, W. F. 1970. The evolutionary significance of recombination in prokaryotes. Soc. Gen. Microb. Symp., 20:279–294.

BRADSHAW, A. D. 1972. Some of the evolutionary consequences of being a plant. Evol. Biol., 5:25–47.

BREMER, K., AND H.-E. WANNTORP. 1979a. Geographic populations or biological species in phylogeny reconstruction? Syst. Zool., 28:220–224.

BREMER, K., AND H.-E. WANNTORP. 1979b. Hierarchy and reticulation in systematics. Syst. Zool., 28:624–627.

BURGER, W. C. 1975. The species concept in *Quercus*. Taxon, 24:45–50.

CLAUSEN, J., D. D. KECK, AND W. M. HIESEY. 1939. The concept of species based on experiment. Amer. J. Bot., 26:103–106.

CLAUSEN, J., D. D. KECK, AND W. M. HIESEY. 1940. Experimental studies on the nature of species. I. The effect of varied environments on Western North American plants. Carnegie Inst. Wash., Publ. No. 520, 452 pp.

CLÉMENÇON, H. (ed.) 1977. The species concept in Hymenomycetes. J. Cramer, Vaduz, Liechtenstein, 444 pp.

COWAN, S. T. 1962. The microbial species – a macromyth? Soc. Gen. Microb. Symp., 12:433–455.

CRONQUIST, A. 1978. Once again, what is a species? Pp. 3–20, *in* Biosystematics in agriculture (J. A. Romberger, ed.). Allanheld & Osmun, Montclair, N.J., 340 pp.

DAVIS, P. H., AND V. H. HEYWOOD. 1963. Principles of angiosperm taxonomy. Oliver and Boyd, Edinburgh, 556 pp.

DOBZHANSKY, T. 1937. Genetics and the origin of species. Columbia Univ. Press, New York, 364 pp.

DOBZHANSKY, T. 1972. Species of *Drosophila*. Science, 177:664–669.

EHRLICH, P. R., AND P. H. RAVEN. 1969. Differentiation of populations. Science, 165:1228–1232.

ELDREDGE, N., AND J. CRACRAFT. 1980. Phylogenetic patterns and the evolutionary process. Columbia Univ. Press, New York, 349 pp.

ELDREDGE, N., AND S. J. GOULD. 1972. Punctuated equilibria: an alternative to phyletic gradualism. Pp. 82–115, *in* Models in paleobiology (T. J. M. Schopf, ed.). Freeman, Cooper and Co., San Francisco, 250 pp.

ENDLER, J. A. 1973. Gene flow and population differentiation. Science, 179:243–250.

FEYERABEND, P. 1970. Consolations for the specialist. Pp. 197–230, *in* Criticism and the growth of knowledge (I. Lakatos and A. Musgrove, eds.). Cambridge Univ. Press, London, 282 pp.

GHISELIN, M. T. 1974. A radical solution to the species problem. Syst. Zool., 23:536–544.

GHISELIN, M. T. 1981. The metaphysics of phylogeny. [Review of Eldredge, N., and J. Cracraft. 1980. Phylogenetic patterns and the evolutionary process.] Paleobiology, 7:139–143.

GOULD, S. J. 1979. A quahog is a quahog. Nat. Hist., 88:18–26.

GOULD, S. J. 1982. Darwinism and the expansion of evolutionary theory. Science, 216:380–387.

GOULD, S. J., AND R. C. LEWONTIN. 1979. The spandrels of San Marco and the Panglossian paradigm: a critique of the adaptionist programme. Proc. Roy. Soc. Lond. (B), 205:581–598.

GRANT, V. 1957. The plant species in theory and practice. Pp. 39–80, *in* The species problem (E. Mayr, ed.). American Association for the Advancement of Science Publ., Washington, D.C., 50 395 pp.

GRANT, V. 1964. The biological composition of a taxonomic species in *Gilia*. Adv. Genet., 12:281–328.

GRANT, V. 1971. Plant speciation. First edition. Columbia Univ. Press, New York, 435 pp.

GRANT, V. 1980. Gene flow and the homogeneity of species populations. Biol. Zbl., 99:157–169.

GRANT, V. 1981. Plant speciation. Second edition. Columbia Univ. Press, New York, 563 pp.

HENNIG, W. 1966. Phylogenetic systematics. Univ. Illinois Press, Urbana, 111., 263 pp.

HILL, C. R., AND P. R. CRANE. 1982. Evolutionary cladistics and the origin of angiosperms. Pp. 269–361, *in* Problems of phylogenetic reconstruction (K. A. Joyse and A. E. Friday, eds.). Systematics Association Special Volume No. 21. Academic Press, London and New York, 442 pp.

HOLMES, E. B. 1980. Reconsideration of some systematic concepts and terms. Evol. Theory, 5: 35–87.

HULL, D. L. 1965. The effect of essentialism on taxonomy – two thousand years of stasis (II). British J. Phil. Sci., 16:1–18.

HULL, D. L. 1968. The operational imperative: sense and nonsense in operationism. Syst. Zool., 17:438–457.

HULL, D. L. 1970. Contemporary systematic philosophies. Ann. Rev. Ecol. Syst., 1:19–54.

HULL, D. L. 1976. Are species really individuals? Syst. Zool., 25:174–191.

HULL, D. L. 1978. A matter of individuality. Phil. Sci., 45:335–360.

HULL, D. L. 1980. Individuality and selection. Ann. Rev. Ecol. Syst., 11:311–332.

JACKSON, J. F., AND J. A. POUND. 1979. Comments on assessing the dedifferentiating effect of gene flow. Syst. Zool., 28:78–85.

JAIN, S. K., AND A. D. BRADSHAW. 1966. Evolutionary divergence among adjacent plant populations. I. The evidence and its theoretical analysis. Heredity, 21:407–441.

KIANG, Y. T. 1982. Local differentiation of *Anthoxanthum odoratum* L. populations on roadsides. Amer. Midi. Nat., 107:340–350.

KUHN, T. S. 1970. The structure of scientific revolutions. Second enlarged edition. Univ. Chicago Press, Chicago, 210 pp.

LANDE, R. 1980. Genetic variation and phenotypic evolution during allopatric speciation. Amer. Nat., 116:463–479.

LEVIN, D. A. 1978. The origin of isolating mechanisms in flowering plants. Evol. Biol., 11:185–317.

LEVIN, D. A. 1979. The nature of plant species. Science, 204:381–384.

LEVIN, D. A. 1981. Dispersal versus gene flow in plants. Ann. Missouri Bot. Gard., 68:233–253.

LEVIN, D. A., AND H. W. KERSTER. 1974. Gene flow in seed plants. Evol. Biol., 7:139–220.

LEWONTIN, R. C. 1974. The genetic basis of evolutionary change. Columbia Univ. Press, New York, 346 pp.

MAYR, E. 1942. Systematics and the origin of species: from the viewpoint of a zoologist. Columbia Univ. Press, New York, 334 pp.

MAYR, E. 1957. Species concepts and definitions. Pp. 1–22, *in* The species problem (E. Mayr, ed.). American Association for the Advancement of Science Publ., Washington, D.C., 50 395 pp.

MAYR, E. 1969. Principles of systematic zoology. McGraw-Hill Book Co., New York, 428 pp.

MAYR, E. 1970. Populations, species, and evolution. Harvard Univ. Press, Cambridge, Mass., 453 pp.

MAYR, E. 1982. The growth of biological thought. Harvard Univ. Press, Cambridge, Mass., 974 pp.

MOONEY, H. A., AND W. D. BILLINGS. 1961. Comparative physiological ecology of arctic and alpine populations of *Oxyria digyna*. Ecol. Mono-gr., 31:1–29.

NELSON, G., AND N. PLATNICK. 1981. Systematics and biogeography: cladistics and vicariance. Columbia Univ. Press, New York, 567 pp.

ORNDUFF, R. 1969. Reproductive biology in relation to systematics. Taxon, 18:121–133.

RAVEN, P. H. 1976. Systematics and plant population biology. Syst. Bot., 1:284–316.

RICKLEFS, R. E. 1979. Ecology. Second edition. Chiron Press, New York, 966 pp.

ROSEN, D. E. 1978. Vicariant patterns and historical explanations in biogeography. Syst. Zool., 27: 159–188.

ROSEN, D. E. 1979. Fishes from the uplands and intermontane basins of Guatemala: revisionary studies and comparative geography. Bull. Amer. Mus. Nat. Hist., 162:267–376.

SIMPSON, G. G. 1961. Principles of animal taxonomy. Columbia Univ. Press, New York, 247 pp.

SMALL, E. 1971. The evolution of reproductive isolation in *Clarkia*, section *Myxocarpa*. Evolution, 25:330–346.

SOKAL, R. R. 1973. The species problem reconsidered. Syst. Zool., 22:360–374.

SOKAL, R. R., AND T. J. CROVELLO. 1970. The biological species concept: a critical evaluation. Amer. Nat., 104:127–153.

STACE, C. A. 1978. Breeding systems, variation patterns and species delimitation. Pp. 57–78, *in* Essays in plant taxonomy (H. E. Street, ed.). Academic Press, New York, 304 pp.

STANLEY, S. M. 1975. A theory of evolution above the species level. Proc. Nat. Acad. Sci. U.S.A., 72: 646–650.

STEBBINS, G. L. 1950. Variation and evolution in plants. Columbia Univ. Press, New York, 643 pp.

STEBBINS, G. L. 1979. Fifty years of plant evolution. Pp. 18–41, *in* Topics in plant population biology (O. T. Solbrig, S. Jain, G. B. Johnson, and P. H. Raven, eds.). Columbia Univ. Press, New York, 589 pp.

STEVENS, P. F. 1980a. A revision of the Old World species of *Calophyllum* L. (Guttiferae). J. Arnold Arb., 61:117–699.

STEVENS, P. F. 1980b. Evolutionary polarity of character states. Ann. Rev. Ecol. Syst., 11:333–358.

TURESSON, G. 1922a. The species and the variety as ecological units. Hereditas, 3:100–113.

TURRESON, G. 1922b. The genotypical response of the plant species to the habitat. Hereditas, 3: 211–350.

VAN VALEN, L. 1976. Ecological species, multispecies, and oaks. Taxon, 25:233–239.

WAKE, D. B. 1980. A view of evolution [Review of Eldredge, N., and J. Cracraft. 1980. Phylogenetic patterns and the evolutionary process.]. Science, 210:1239–1240.

WHITE, M. J. D. 1978. Modes of speciation. W. H. Freeman and Co., San Francisco, 455 pp.

WILEY, E. O. 1978. The evolutionary species concept reconsidered. Syst. Zool., 27:17–26.

WILEY, E. O. 1980. Is the evolutionary species fiction? – A consideration of classes, individuals, and historical entities. Syst. Zool., 29:76–80.

WILEY, E. O. 1981. Phylogenetics: the theory and practice of phylogenetic systematics. John Wiley, New York, 439 pp.

WILEY, E. O., AND D. R. BROOKS. 1982. Victims of history – a nonequilibrium approach to evolution. Syst. Zool., 31:1–24.

The Morphological, Developmental, and Phylogenetic Basis of Species Concepts in Bryophytes[7]

Abstract

This paper examines the theoretical and practical status of species relative to two major issues: the recognition of the importance of epigenetic constraints in evolution and the rise of Hennigian phylogenetic systematics (cladistics). Theories advanced to explain the origin and maintenance of basic morphological clusters of organisms (species) have usually involved two main classes of causal factors: selection (ecological constraints) and gene flow (breeding barriers vs. the integrating effect of gene exchange). However, in many plants, non-correspondence of patterns of discontinuities among basic morphological, ecological, and breeding groups has been noted. The "biological species concept" is flawed because it is biased towards explanations at the genetic level. A third class of causal factors (epigenetic constraints) has come into favor as an explanation for the distinctness of higher-level morphological clusters, but the relevance of epigenetic factors as primary constraints on morphological variation at the species level remains to be examined. A phylogenetic species concept is advocated, which views species as monophyletic groups of organisms, the smallest such groups recognized in a formal classification. Assignment of species rank to a particular group should depend on the causal factors acting to maintain that group as an independent lineage. Epigenetic constraints may prove to be the most important factor producing and maintaining species lineages. Bryophytes are useful organisms for investigating this question because they are readily manipulated under experimental conditions, both sexual and asexual species exist, and a diversity of ecological and geographic specificities are known.

What are basic kinds in systematic and evolutionary biology? Since the Darwinian Revolution, biologists have been struggling to arrive at a concept of species in order to explain morphological discontinuities in terms of the dynamic process

---

[7] B.D. Mishler. 1985. The morphological, developmental, and phylogenetic basis of species concepts in bryophytes. The Bryologist 88: 207–214. [reprinted by permission]

of evolution. This debate has not been resolved. Botanists and zoologists, phyleticists and pheneticists, paleontologists and population geneticists, philosophers and practitioners have yet to reach a consensus.

Simple observation shows that the diversity of the natural world is not continuous. Organic variation usually falls into clusters or nodes, which are often structured hierarchically into progressively more inclusive clusters. The central problem of systematic biology is to document these patterns; the central problem of evolutionary biology is to determine what processes are responsible for stasis or change in these nodes. Clearly, the units recognized by systematists must be the units involved in processes studied by evolutionists if the work of these two fields is to be mutually relevant. Therefore, it is important to examine carefully the status of species, a category often held to be unique and fundamental by both systematists and evolutionists.

A diversity of views on the meaning of species exists (for reviews see Mayr 1957; Wiley 1978). Yet it seems that the prevailing species concept among systematists, ecologists, and evolutionists (especially zoologists) is some version of the biological/evolutionary species concept (Simpson 1961; Mayr 1970) or the "species as individual" concept as developed by Ghiselin (1974) and Hull (1976, 1980). In this view species taxa are seen as fundamentally and universally different from taxa at all other levels: real, genetically integrated, cohesive, and comparable units of evolution.

The acceptance of the biological or evolutionary species concept has certainly not been universal among botanists (Mishler & Donoghue 1982). Some botanists, especially cytologists (e.g., Stebbins 1950, 1979) advocate a biological species concept based on discontinuities in gene flow. However, many botanists favor a morphological (i.e., phenetic) species concept (e.g., Cronquist 1978; Levin 1979). However, neither of these concepts is appropriate for recognizing theoretically meaningful taxa.

Mishler and Donoghue (1982) presented empirical and theoretical arguments to indicate that current biological or evolutionary species concepts are oversimplified (see also Donoghue, this symposium). They concluded that no single and universal basic evolutionary unit exists and that in most cases species taxa have no special reality in nature. Furthermore, criteria for what constitutes an "important" discontinuity appear to depend on differences in the biology of different groups of organisms. Causal factors responsible for the existence of morphologically distinct species seem to be fundamentally different in different groups. It was suggested that a more pluralistic concept of species is needed to reflect adequately the variety and complexity of "species situations."

In the present paper, I attempt to weave together two theoretical strands. The first relates to the general debate in systematics over what taxonomic names should represent. Even among Hennigian phylogenetic systematists a diversity of species concepts has been or is being espoused. While this school of thought is presenting a strong challenge to traditional "evolutionary systematics," there has been inadequate discussion of the fundamental taxonomic level of species. The second strand involves the debate over the relative importance of various evolutionary forces acting to constrain morphological variation over time. Growing realization of the importance of epigenetic constraints in evolution is another element in an increasingly coherent challenge to the neo-Darwinian "modern synthesis." However, whether epigenetic factors act as constraints on morphological variation at the species level remains to be examined.

This discussion will of necessity be generalized because the application and meaning of the species category remain problematic in all major groups of organisms. However, in keeping with the theme of the symposium, I will refer primarily to bryophytes (especially to the moss *Tortula*) for examples. What is special about the species situation in bryophytes? Conversely, how can bryophytes illustrate issues of general biological interest with respect to the species problem?

The Species Category and Hennigian Phylogenetic Systematics: Ontology and Phylogeny

The past two decades have seen the rise of Hennigian phylogenetic systematics or cladistics. This method arose as a reaction to the traditional, subjective "evolutionary systematics" and to the more recent indiscriminate grouping by overall similarity advocated by the numerical phenetics school. The application of Hennigian cladistics to bryology was discussed by Mishler and Churchill (1984); see Wiley (1981) for further details.

Two important ideas contributed by Hennig (1966) were the recognition of the importance of using shared derived characters for reconstructing phylogeny and the restriction of the concept of monophyly to those groups that contain all and only descendants of a common ancestor.

Within the Hennigian school a diversity of species concepts are advocated, none of which fully deals with the application of Hennigian ideas such as monophyly to the species category. Hennig (1966) advocated the biological species concept in his work. As expressed by Mayr (1970), biological species "are groups of interbreeding natural populations that are reproductively isolated from other such groups."

Nelson and Platnick (1981) devoted only a few paragraphs to the species category. They define species as "simply the smallest detected samples of self-perpetuating organisms that have unique sets of characters," then go on to stress that a species need not have any single unique trait – only a unique, diagnosable combination of characters. They make a distinction between "species" and "groups of species." As Platnick (1985) has elaborated, the latter does need to have apomorphies. However, a basic concept (such as "species") that fills an ontological role in a theoretical system cannot be chosen arbitrarily and independently of the theoretical system in which it occurs. Since Nelson and Platnick are not enthusiastic about current evolutionary theory, it is perhaps understandable that they do not frame and justify their species concept in terms of that theory. However, their concept seems inadequately justified even in terms of the theoretical system in which they work. There is no explicit connection to the general concept of taxa as lineages, and there is also nothing to distinguish their concept from those proposed by pheneticists. The use of phenetic species in cladistic and biogeographic analyses is an important internal conflict. A phenetic species concept at best can serve as a first approximation of basal cladistic units in a phylogenetic analysis, but the concept of species needed for phylogenetic systematics must be richer in content.

Rosen (1979) has defined species as a "population or group of populations defined by one or more apomorphous features … the smallest natural aggregation of individuals with a specifiable geographic integrity that can be defined by any current set of analytical techniques." This definition, in contrast to that of Nelson and Platnick, is consistent with the systematic and biogeographic theories of the author. However, it is atomistic, tending towards the recognition of individual organisms or family groups as Linnaean species (Hill & Crane 1982; Mishler & Donoghue 1982). It is important to recognize that concepts of phylogenetic taxa (including species) must include two components, a grouping criterion (e.g., monophyly) and a ranking criterion (see Donoghue, this symposium). Rosen's definition has a criterion that can be used to group organisms, but the criterion for ranking is vague. I will return later to this distinction and to the question of just what these ranking criteria should be in phylogenetic systematics.

Eldredge and Cracraft (1980) defined species as: "a diagnosable cluster of individuals within which there is a parental pattern of ancestry and descent, beyond which there is not, and which exhibits a pattern of phylogenetic ancestry and descent among units of like kind." While this definition out of context may not be clear, since all would accept that there is a pattern of ancestry and descent beyond the species level, reference to the discussion in Eldredge and Cracraft (1980) will show that the concept being advocated is more or less the standard biological species concept. However, Cracraft (1982) has more recently presented a view of species similar to that of Nelson and Platnick (1981).

As a modification of Simpson's (1961) evolutionary species concept Wiley (1981:25) has presented the following species definition: "a single lineage of ancestral descendant populations of organisms which maintains its identity from other such lineages and which has its own evolutionary tendencies and historical fate." I find this concept to be the most compelling so far advocated by cladists because it provides an explicit statement of species as lineages. However, this concept is similar to the biological species concept in its usual emphasis on genetic discontinuity and in that it views the species category as unique and fundamental (a view that I argue is not generally the case).

Taken as a whole, the various concepts of species held by cladists span almost the whole range of diversity of species concepts held by systematists generally. However, these cladistic species concepts have features in common. All (except for Rosen's concept) make a clear distinction between species and groups of species (i.e., higher taxa) and do not apply an unequivocal, cladistic concept of monophyly (i.e., the recognition of monophyly by apomorphies) to the species level. This distinction is also made by Hill and Crane (1982) who contrast cladistic species (as evidenced by apomorphies) with "real" species (as evidenced by genetic isolation) and point out that there is no necessary congruence between the two.

There are also substantial differences among the species concepts discussed above. What species concept is adopted matters theoretically because of the central role of the species concept and of species in biology and practically because systematists with differing views of species will often treat the same situation in different ways. Terms such as "species" have a necessary relationship to theories in which they are embedded (Hull 1981). As Hull (1968, 1981), Beatty

(1982), Lewontin (1974), Hennig (1966), and others have pointed out, progress in understanding the world comes as we build an elaborate theoretical structure by iteratively matching together lower-level theories (such as species concepts and classifications) with higher-level theories (such as evolutionary theory). The question of definition or meaning of a basic term such as species is thus embedded in both lower-level and higher-level causal theories. Empirical considerations enter in the low-level matching of observation and theoretical concept; ontological considerations enter in because of the function of such basic terms in higher-level explanatory theories.

Despite the many problems with details, the theory of evolution (descent with modification) unquestionably remains the primary overarching theoretical structure for comparative biology and provides both the justification and motivation for the development of phylogenetic systematics. Concepts of species must be evaluated in terms of the role these concepts play in evolutionary theory as well as by their fit to observed patterns of variation among organisms.

The evolutionary species concept as developed by Wiley and others fails to fit observed empirical patterns in many groups of organisms (Mishler & Donoghue 1982). There are two main conclusions drawn by Mishler and Donoghue which, if true, have important implications for phylogenetic systematics. The first is the notion that species often have no special reality as compared to taxa at other levels within a clade. The second is that the causal factors responsible for differentiation at the species level seem to be fundamentally different in different groups.

The first conclusion conflicts with the current distinction between "species" and "groups of species," which figures importantly in cladistic concepts of monophyly, ancestor-descendant relationships, and the supposed difference between cladograms and trees. Since (contrary to Wiley 1981) species cannot be taken as a priori monophyletic, cladistic theory must expand and develop meaningful concepts of monophyly at the species level. The task of developing a revised definition of monophyly that is applicable to the species category (which is necessary because Hennigian monophyly is traditionally defined with reference to origin of a taxon in a single species) is beyond the scope of the present paper. In short, monophyly should be viewed as involving a single origin of a taxon. In the case of species, this means single origin in one population or even one individual organism.

Particularly controversial situations that must be faced by a strict phylogenetic species concept are the apparently frequent cases of incongruence between cladistically discrete groups and genetically discrete groups (biological species) (Bremer & Wanntorp 1979; Hill & Crane 1982). As pointed out by Rosen (1979), biological species may often be united only by a plesiomorphy, reproductive compatibility. A solution favored in these situations by Wiley (1981, and pers. comm.) and Hill and Crane (1982) is to name as species the genetically interconnected groups. In my opinion, however, this would obscure the important pattern. The fact that cladistic structure can be observed in the face of potential or actual gene flow is an indication that other causal factors are operating and a good reason to name as species the cladistically distinct groups. I favor applying Hennigian principles of classification to the use of

Linnaean binomials, reserving the formal taxon, species, for minimally "important" phylogenetic units as evidenced by apomorphies. In this context "important" refers to ranking criteria.

The generally unrecognized distinction between the grouping and ranking components of a species concept is critical (see also discussion by Donoghue, this symposium). All species concepts have both components. For example, in the biological species concept, the grouping criterion is having the ability to interbreed; the ranking criterion is being the most inclusive group having that ability. Even when monophyletic groups (sensu Hennig) are delimited the problem of ranking remains because monophyletic groups can be found at many levels within a clade. Species ranking criteria could include morphological gap size (i.e., morphological distinctiveness), ecological or geographical criteria, degree of intersterility, and possibly others. The general problem of ranking is presently unresolved. I contend that an absolute and universally applicable criterion is not possible and that criteria will have to be developed on a group by group basis with reference to causal factors that are important in maintaining lineages in particular groups. The grouping concept, monophyly, should be retained as a universal component of the species concept until it can be shown empirically that other concepts of grouping are needed (which may indeed be the case in groups such as bacteria).

An Epigenetic View of the Species Category: Ontogeny and Plurality

A renewal of interest in the relationship between development and evolution has taken place in evolutionary biology. Darwin's mysterious "laws of correlation of characters" did not figure prominently in the modern synthesis of evolutionary theory fashioned in the 1930s and 40s despite Waddington's (1957) striking analogy of a developmental landscape. [See Gould (1980) for a general discussion of current challenges to the "modern synthesis," which hardened in the late 1940s and 50s into a strongly reductionistic, genetically based theory.]

Of particular relevance here are recent attempts to discover and define developmental or epigenetic constraints on patterns of morphological form (Ho & Saunders 1979; Alberch 1980, 1982; Rachootin & Thomson 1981; Oster & Alberch 1982). The general theme of these investigations is the existence of complex epigenetic programs that translate the genotype into the phenotype. These epigenetic rules ensure that a one-to-one matching between a particular sequence of DNA and a particular phenotypic feature is unusual. Thus, certain phenotypes are prohibited or are unlikely, and continuous genetic change can result in discontinuous epigenetic changes.

The possible role of epigenetic factors as primary constraints on morphological variation at the species level has only been mentioned in passing (Eldredge & Gould 1972). In my opinion, an important internal conflict is created when workers such as Gould (1979) and Eldredge and Cracraft (1980) defend the strongly reductionistic, genetically based biological species concept at the same time they are advocating a critical reexamination of the modem synthesis. The newly developing macroevolutionary theory relies strongly on a hierarchical view of evolution (Gould 1980; Eldredge & Cracraft 1980), but the biological species concept is not necessary (in fact, it may even be harmful) for this view.

The debate about the differentiation or stasis of populations and species has so far primarily been a comparison of two classes of causal factors, gene flow and selection (Ehrlich & Raven 1969; Jackson & Pounds 1979; Ehrlich & White 1980; Pounds & Jackson 1982). Neither position seems to provide a satisfactory explanation for many situations in plants. As pointed out by Grant (1980:167), "the homogeneity of species is due more to descent from a common ancestor than to gene exchange across significant parts of the species-area." Of course, this well-justified conclusion by itself does not provide an explanation for the homogeneity of species; the addition of a third class of causal factors, epigenetic constraints, is necessary.

It is well known that morphological variation is often poorly connected to genetic variation in plants. In some cases small genetic changes can lead to major morphological changes (Gottlieb 1984); in other cases, large genetic changes result in little or no morphological difference. Examples of the latter are provided by single morphological species with a variable chromosome number (e.g., *Sedum moranense*, Uhl 1983; and *Tortula princeps*, Steere, Anderson & Bryan 1954) or the widely recognized existence of high levels of genetic polymorphism (as revealed by electrophoresis) within and between single populations (Lewontin 1974; reviewed for bryophytes by Wyatt 1982).

Examples of noncorrespondence between morphological, breeding, and ecological discontinuities in plants are numerous (see literature cited in Mishler & Donoghue 1982), but one example from bryophytes is instructive. My studies of the moss *Tortula* (Mishler 1984) illustrated the sorts of noncorrespondence that have caused dissatisfaction with the prevailing view of gene flow and selection as the only major causal factors in the homogeneity or divergence of species and have prompted a search for another explanation.

Tortula contains both sexual and asexual species, a spectrum of sexual conditions from dioicy to synoicy in the sexual species, and several types of specialized asexual propagules. *Tortula* includes species that are narrowly endemic, others that are cosmopolitan, and some that are intermediate. Ecologically, both specialized and generalized species are represented. This diversity can allow comparative study of topics of current interest, such as punctuated morphological change and the relative importance of gene flow, selection, developmental constraints, and historical factors in producing and maintaining discontinuities in morphological variation.

The 21 species of *Tortula* recognized for Mexico (Mishler 1994) do not form a monophyletic group, but they represent a cross-section of the diversity of species within the genus. These species can be divided into four discrete categories: sporophytes always present (7 species); sporophytes infrequent but produced in a roughly equal frequency throughout the range of the species (6 species); sporophytes infrequent but produced only in a small part of the total range, the species thus asexual throughout most of their ranges (5 species); and sporophytes absent and the species thus apparently wholly asexual (3 species). The lack of certain expected correlations with the frequency of sporophyte production (and hence the opportunity for reproducing via sexually produced propagules) is striking. The size of a species' geographic range or its frequency of occurrence within it (as judged from herbarium specimens and my extensive field studies in Mexico) is not correlated with the frequency of sporophyte formation. Furthermore, the distinctiveness of a particular species (i.e., the possession of a set of unique characters) is not correlated with either sporophyte frequency or the occurrence of specialized asexual propagules.

The evident importance of asexual reproduction in many species of *Tortula* and the occurrence of widespread species that are nonetheless constant morphologically seem to eliminate gene flow as a likely integrating factor. Stabilizing selection may be acting to maintain the morphologically distinctive features of each species of *Tortula* in Mexico. But this explanation seems unlikely because while each species has a rather restricted set of preferred habitats, the various species overlap in their preferences, and a single species may have a considerable ecological amplitude without concomitant morphological variation. For example, *T. quitoensis* Taylor occurs disjunctively from the northern Andes to the subalpine and alpine zones of the high volcanoes of Mexico. Plants of this species produce sporophytes infrequently but asexual reproduction is possible via the easily detachable leaves. The species maintains its distinctive morphology while occurring on soil of disturbed trail banks, on the bark of trees in shaded situations, and on moist rocks near streams – a broad habitat preference for any moss, let alone a species with a restricted disjunct distribution. The question remains: if neither patterns of gene flow nor a specifiable ecological "niche" appear to cause the morphological coherence of species, what does?

While not yet well-known in a mechanistic sense, a compelling answer to this question is the notion of epigenetic constraints on species. A species in this sense is a monophyletic lineage (recognized as such by the possession of a set of unique characters), buffered against small amounts of gene exchange because of the action of homeostatic epigenetic programs. Such a lineage might be fuzzy at the edges, but still be recognizable as a lineage, just as the Gulf Stream is recognizably a current (and has causal efficacy for the climate of Britain) despite the loss and gain of some water molecules along the way. Biological examples are many North American oak species, which seem to maintain their integrity despite some inter-crossing and apparent ecological overlap.

Just how much genetic intermixing causes a blending of two lineages (as defined above) appears to depend on what other processes are operating in a particular case. This returns the discussion to pluralism and the distinction between grouping and ranking criteria. The grouping criterion in the species concept advocated here remains monophyly. Various causal factors (epigenetic constraints, discontinuities in gene flow, or stabilizing selection) may figure in the ranking criterion needed in particular cases. This distinction is important because, as noted by several authors, the action of epigenetic constraints can lead to a strong bias towards parallelism. In other words, organisms with similar developmental programs may independently undergo channelized morphological changes, thus independently gaining similar phenotypes. However, using the grouping criterion of monophyly requires one to make every attempt to distinguish polyphyletic from monophyletic groups and formally name only the latter.

The concept of epigenetic constraints remains suggestive and is imperfectly known, but it is an exciting field for future research. The full importance of such constraints will only become clear with a greater understanding of development than is currently present. Further study is needed to discover patterns of continuity and discontinuity in the transition from genes through epigenetic programs to phenotypes.

Bryophytes appear to be an advantageous group of organisms for experimentally examining these questions because of the ease with which clones can be replicated and because a large number of plants can be cultured in a small space. Their development is rather simple and yet a considerable morphological, reproductive, and ecological diversity is present. The work of Basile and Basile (1983, 1984) has demonstrated the feasibility of documenting and experimentally manipulating morphogenetic processes in liverworts. Wettstein's early experimental work with mosses (reviewed by Anderson 1963) likewise shows great promise.

Implications of a Phylogenetic Concept of Species

The phylogenetic concept of species developed in this paper and in Donoghue (this symposium) has both theoretical and practical implications. Usefulness is often taken to be an important virtue of a classification. As discussed above, an important part of the usefulness of taxa (including species) resides in their ontological role in causal theories meant to explain them. Taxa that are monophyletic in the Hennigian sense are necessary for more than purposes of phylogenetic reconstruction. Species that are historical (i.e., monophyletic) entities are necessary if meaningful studies of biogeography, speciation, historical ecology, or comparative morphology are to be carried out. Phenetic species are unsuitable for these purposes because grouping by overall similarity will only sometimes recognize monophyletic lineages.

The phylogenetic species concept seems to fit the theoretical uses to which botanists have always applied their species. Traditional studies of phytogeography, speciation, and evolution have treated species as "things" (i.e., historical entities with many of the attributes of individuals in the philosophical sense; Ghiselin 1974; Hull 1976, 1980). The concept of species advocated here merely provides an explicit connection between the concept and the uses to which it is put.

In a practical sense, usefulness has been claimed for phenetic taxa (including species, e.g., Levin 1979) without challenge until recently. As discussed by Farris (1983), many measures of information content are satisfied better by cladistic classifications than by phenetic ones. Therefore, not only do cladistically defined taxa have the primary benefit of theoretical meaningfulness, they are also more informative about known and predicted biological information than phenetically defined taxa. Thus phylogenetic species are more practical in the sense botanists have used the term when advocating phenetic concepts.

What are the taxonomic implications of the phylogenetic species concept advocated here? How does a practicing systematist proceed in a group such as the bryophytes that is little known biologically? One proceeds using the method of reciprocal illumination (Hennig 1966) to study morphology and produce a hypothesis of basic monophyletic groups (i.e., species) evidenced as groups by apparent apomorphies and perhaps initially ranked as species only because of the size of morphological gaps. One then progressively adjusts the circumscription of species as more becomes known about gene flow, ecological relationships, and the genetic and epigenetic basis of characters.

The astute bryological taxonomist will realize by this point that many species especially in taxonomically difficult groups such as the *Tortula ruralis* complex, *Bryum*, or *Brachythecium* cannot currently be defined by the possession of autapomorphies. However, considerable lumping may be required to recognize species in these groups that are characterized by autapomorphies. This may be precisely why these groups are taxonomically difficult; current taxonomic work is below the level at which clearly monophyletic lineages can be recognized. An analogy may be in order: the situation in such difficult groups is rather like trying to read a book with a microscope – the very quality of "letterness" disappears into blobs of ink and fibers of paper. Meaningful letters come into focus only when examined with less magnification.

If it is accepted that a species should be named formally only when a hypothesis is made that it is a historically discrete, monophyletic entity, then specific and infraspecific taxa characterized only by shared primitive characters (plesiomorphies) are not satisfactory. In such cases, attempts should be made to find characters that either demonstrate the monophyly of the taxon or conversely demonstrate how the taxon could be broken up into monophyletic groups. Such attempts may fail, since "paraphyletic" speciation is a real possibility, as in the case of a widespread species that gives rise to other species through peripheral isolation (Bremer & Wanntorp 1979; Mishler & Donoghue 1982; Donoghue this symposium). Ackery and Vane-Wright (1984) and Donoghue (this symposium) discussed this important problem in the application of monophyly to species. Donoghue noted that some species have an intermediate status between monophyly and paraphyly, in having no autapomorphies and yet with no characters that demonstrate paraphyly. Such a group of uncertain monophyly can be recognized formally under the phylogenetic species concept advocated here (as long as the group is not demonstrably paraphyletic), but should be distinguished in some way from demonstrably monophyletic species. [See further discussion of this topic by Donoghue, this symposium.]

Clearly, the recognition of polyphyletic taxa at or below the level of species, as for example in the naming of taxa on the basis of a few, environmentally labile characters or a simple genetic polymorphism, is unsatisfactory. Such ecological modifications (or even minor genetical variants) are likely to appear independently many times within a single, monophyletic species and should be given at most only informal recognition.

Examples of these considerations are provided by biosystematic studies of the *Tortula ruralis* complex (Mishler 1984, 1985). *Tortula ruralis* itself as currently circumscribed is problematical because it appears to have no unique characters to define it as a species, there are elements within it that appear to be defined by unique characters, and some elements of it share possibly unique characters with other species. This "species" is thus likely to be paraphyletic. Examples of currently recognized species in the complex that are likely to be polyphyletic are *T. ruraliformis* and *T. intermedia*. Preliminary culture studies of these "species" indicate that characters held to distinguish them are quite plastic phenotypically and thus not reliable indicators of monophyly.

I contend that formal infraspecific names should be subject to the same criteria as species names in order to have a taxonomic system that is completely consistent and theoretically meaningful. The former should be used only if a hypothesis of monophyly is made and if there is some reason why the taxa recognized are not ranked at the species level.

If formal taxonomic names are used exclusively for phylogenetically discrete units at all taxonomic levels, then of course, many other biologically important units (such as physiognomic types, ecological guilds, ecophenotypes, cytotypes) will not be formally named. Nothing needs to be lost, however, and much may be gained. These units can be described informally and placed in overlapping, non-hierarchical arrangements as required. As such they will be clearly distinguished from monophyletic groups that bear formal names.

To summarize, I have advocated the proposition that species should be viewed as monophyletic groups of organisms, recognized as lineages on the morphological basis of the possession of shared, derived characters, and ranked as species because of causal factors (perhaps especially epigenetic constraints) that maintain the lineage as the smallest important monophyletic group recognized in a formal classification. Such a view of species, while "morphological" in an important sense, is different than previous morphological or phenetic species concepts. It can provide species that are theoretically meaningful for biogeographic, ecological, and evolutionary studies and that have the practical virtues of consistency and predictiveness.

ACKNOWLEDGMENTS

John Beatty, Michael Donoghue, and Peter Stevens provided stimuli for the development of many of these ideas. Portions of this paper were also presented at the Third Annual Meeting of the Willi Hennig Society in College Park, Maryland, 1982. I thank Sara and William Fink, David Hull, Norton Miller, Ronald Pursell, and Peter Stevens for comments on versions of the manuscript.

LITERATURE CITED

ACKERY, P. R. & R. I. VANE-WRIGHT. 1984. Milkweed Butterflies: Their Cladistics and Biology. Cornell Univ. Press, Ithaca, New York

ALBERCH, P. 1980. Ontogenesis and morphological diversification. American Zoologist 20: 653–667.

_____. 1982. Developmental constraints in evolutionary processes, pp. 313–332. In J. T. Bonner (ed.), Evolution and Development. Springer-Verlag, Berlin.

ANDERSON, L. E. 1963. Modern species concepts: Mosses. The Bryologist 66: 107–119.

BASILE, D. V. & M. R. BASILE. 1983. Desuppression of leaf primordia of *Plagiochila arctica* (Hepaticae) by ethylene antagonist. Science 220: 1051–1053.
____ & ____. 1984. Probing the evolutionary history of bryophytes experimentally. Journal of the Hattori Botanical Laboratory 55: 173–185.
BEATTY, J. 1982. Classes and cladists. Systematic Zoology 31: 25–34.
BREMER, K. & H.-E. WANNTORP. 1979. Geographic populations or biological species in phylogeny reconstruction? Systematic Zoology 28: 220–224.
CRACRAFT, J. 1982. Are cladistic species biological, evolutionary, or phylogenetic? A discourse on the kinds of cladists. Third Annual Meeting of the Willi Hennig Society, College Park, Maryland.
CRONQUIST, A. 1978. Once again, what is a species?, pp. 3–20. In J. A. Romberger (ed.), Biosystematics in Agriculture. Allanheld & Osmun, Montclair, New Jersey.
EHRLICH, P. R. & P. H. RAVEN. 1969. Differentiation of populations. Science 165: 1228–1232.
____ & R. R. WHITE. 1980. Colorado checkerspot butterflies: Isolation, neutrality, and the biospecies. American Naturalist 115: 328–341.
ELDREDGE, N. & J. CRACRAFT. 1980. Phylogenetic Patterns and the Evolutionary Process. Columbia Univ. Press, New York.
____ & S. J. GOULD. 1972. Punctuated equilibria: An alternative to phyletic gradualism, pp. 82–115. In T. J. M. Schopf (ed.), Models in Paleobiology. Freeman, Cooper & Co., San Francisco.
FARRIS, J, S. 1983. The logical basis of phylogenetic analysis, pp, 7–36. In N. D. Platnick & V. A. Funk (eds.), Advances in Cladistics, Vol. 2, Columbia University Press, New York.
GHISELIN, M. T. 1974. A radical solution to the species problem. Systematic Zoology 23: 536–544.
GOTTLIEB, L. D. 1984. Genetics and morphological evolution in plants, American Naturalist 123: 681–709.
GOULD, S. J. 1979. A quahog is a quahog. Natural History 88: 18–26.
____. 1980. Is a new and general theory of evolution emerging? Paleobiology 6: 119–130.
GRANT, V. 1980. Gene flow and the homogeneity of species populations. Biologisches Zentralblatt 99: 157–169.
HENNIG, W. 1966. Phylogenetic Systematics. Univ. Illinois Press, Urbana, Illinois.
HILL, C. R. & P, R. CRANE. 1982. Evolutionary cladistics and the origin of angiosperms, pp. 269–361. In K. A. Joysey & A. E. Friday (eds.), Problems of Phylogenetic Reconstruction. Academic Press, London.
HO, M. W. & P. T. SAUNDERS. 1979. Beyond neo-Darwinism–an epigenetic approach to evolution. Journal of Theoretical Biology 78: 573–591.
HULL, D. L. 1968. The operational imperative: Sense and nonsense in operationism. Systematic Zoology 17: 438–457.
____. 1976. Are species really individuals? Systematic Zoology 25: 174–191.
____. 1980. Individuality and selection. Annual Review of Ecology and Systematics 11: 311–332.
____. 1981. The principles of biological classification: The use and abuse of philosophy. PSA [Philosophy of Science Association.] 1978(2): 130–153.
JACKSON, J. F. & J. A. POUNDS. 1979. Comments on assessing the de-differentiating effect of gene flow. Systematic Zoology 28: 78–85.
LEVIN, D., A. 1979. The nature of plant species. Science 204: 381–384.
LEWONTIN, R. C. 1974. The Genetic Basis of Evolutionary Change. Columbia University Press, New York.
MAYR, E. 1957. Species concepts and definitions, pp. 1–22. In E. Mayr (ed.), The Species Problem. American Association for the Advancement of Science Publ., Washington, D.C., Vol. 50.

_____ 1970. Populations, Species and Evolution. Harvard University Press, Cambridge, Massachusetts.

MISHLER, B. D. 1984. Systematic studies in the genus, *Tortula* Hedw. (Musci: Pottiaceae). Ph.D. thesis, Harvard University.

_____ (1985). Biosystematic studies of the *Tortula ruralis* complex. I. Variation of taxonomic characters in culture. Journal of the Hattori Botanical Laboratory 58: 225-253.

_____ (1994). *Tortula*. In A. J. Sharp, H. A. Crum, and P. M. Eckel (eds.), The Moss Flora of Mexico. Mem. New York Bot. Gard. 69: 319–352.

_____ & S. P. Churchill. 1984. A cladistic approach to the phylogeny of the "bryophytes." Brittonia 36: 406–424.

_____ & M. J. DONOGHUE. 1982. Species concepts: A case for pluralism. Systematic Zoology 31: 491–503.

NELSON, G, & N. PLATNICK. 1981. Systematics and Biogeography: Cladistics and Vicariance. Columbia University Press, New York.

OSTER, G. & P. ALBERCH. 1982. Evolution and bifurcation of developmental programs. Evolution 36: 444–459.

PLATNICK, N. 1985. Philosophy and the transformation of cladistics revisited. Cladistics 1: 87–94.

POUNDS, J. A. & J. F. JACKSON. 1982. Gene flow and differentiation: The isolated populations of checker-spot butterflies in Colorado. American Naturalist 120: 280–281.

RACHOOTIN, S. P. & K. S. THOMSON. 1981. Epigenetics, paleontology, and evolution, pp. 181–193. In G. G. E. Scudder & J. L. Reveal (eds.), Evolution Today. Carnegie-Mellon University, Pittsburgh.

ROSEN, D. E. 1979. Fishes from the uplands and inter-montane basins of Guatemala: Revisionary studies and comparative geography. Bulletin of the American Museum of Natural History 162: 267–376.

SIMPSON, G. G. 1961. Principles of Animal Taxonomy. Columbia Univ. Press, New York.

STEBBINS, G. L. 1950. Variation and Evolution in Plants. Columbia University Press, New York.

_____. 1979. Fifty years of plant evolution, pp. 18–41. In O. T. Solbrig, S. Jain, G. B. Johnson & P. H. Raven (eds.), Topics in Plant Population Biology. Columbia University Press, New York.

STEERE, W. C., L. E. ANDERSON & V. S. BRYAN. 1954. Chromosome studies on California mosses. Memoirs of the Torrey Botanical Club 20: 1–75.

UHL, C. H. 1983. Chromosomes of Mexican *Sedum*. IV. Heteroploidy in *Sedum moranense*. Rhodora 85: 243–252.

WADDINGTON, C. H. 1957. The Strategy of the Genes. Allen & Unwin, London.

WILEY, E. O. 1978. The evolutionary species concept reconsidered. Systematic Zoology 27: 17–26.

_____. 1981. Phylogenetics: The Theory and Practice of Phylogenetic Systematics. John Wiley, New York.

WYATT, R. 1982. Population ecology of bryophytes. Journal of the Hattori Botanical Laboratory 52: 179–198.

Species and Evolution in Clonal Organisms – Introduction[8]

The species question in biology has a long and tortured past. Despite increasing amounts and types of morphological and genetic data, we seem to be no closer to an answer to "what is a species?" than we were 100 years ago. Yet, although disagreement continues over what species are, many biologists do seem to agree that species do not exist in asexual organisms, at least in an evolutionary or process sense (Eldredge 1985; Ghiselin 1987; Grant 1981; Hull 1987). Despite this widespread consensus, a handful of researchers continue to argue on both empirical and theoretical grounds that species (or units very much like species) do exist in many groups of asexual organisms (Frost and Wright 1988; Holman 1987; Mishler and Brandon 1987; Shaposhnikov 1984; Templeton 1989).

This symposium, entitled "Species and Evolution in Clonal Organisms," was held at the 1988 meeting of the American Society of Zoologists in San Francisco, sponsored by the Division of Systematic Zoology, ASZ, the Society of Systematic Zoology, and the California Academy of Sciences. The overall goal was to take a fresh, empirical look at the nature of species in asexual groups. As organizers, we devised several questions that were circulated beforehand to invited speakers. These included: How often do clonal organisms form discrete species? In cases where they do, what provides the cohesive force for maintaining their discreteness? How is speciation accomplished in clonal organisms? Can study of clonal organisms provide evidence that a process other than gene flow is fundamental in maintaining species even in sexual organisms?

We selected speakers representing a range of different organisms with different reproductive biologies and frequencies of asexual reproduction, including both plants and animals. The speakers at the symposium included: B. D. Mishler (Duke University) on mosses, D. R. Farrar (Iowa State University) on ferns, E. Fenster and U. Sorhannus (Queens College, SUNY) on diatoms, E. Kellogg (Harvard University) on grasses, C. S. Campbell (University of Maine) and T. A. Dickinson (Royal Ontario Museum) on trees and shrubs in the rose family, J. Moore (Colorado State University) on parasitic helminths, B. Willis (James Cook University, Australia) on corals, A. F. Budd (University of Iowa) on corals, and A. H. Cheetham and J. B. C. Jackson (Smithsonian Institution) on bryozoans. The conclusion of the symposium featured a lively and prolonged roundtable discussion (see Summary and Discussion for a synopsis of the roundtable discussion and a list of titles).

[8] B.D. Mishler and A.F. Budd. 1990. Species and evolution in clonal organisms – introduction. Systematic Botany 15: 79–85. [reprinted by permission]

Six papers presented in the symposium follow in this issue, bolstered by the addition of material derived from peer reviews plus discussion at the symposium itself. With four of the six papers dealing with plants, we feel that *Systematic Botany* is an appropriate outlet for this topic of general biological interest. The goal of this introduction is to set the stage by introducing three major issues that are common to all papers. We will first briefly review competing species concepts, and the expected patterns of species distinctness. We will then consider mechanisms possibly responsible for causing integration and cohesion of species. Finally, we will discuss what "speciation" might be under these different concepts of species and under various possible integrating mechanisms.

Species Concepts

Many different concepts of species have been and are held by biologists. This conceptual diversity can be reduced for purposes of discussion by placing these concepts under four headings, based on the way in which the nature of groupings of organisms is viewed under a particular concept (for another recent taxonomy of species concepts see Lidén and Oxelman 1989).

Species as Breeding Groups

The prevailing view of species among biologists at large (if perhaps not among systematists per se) is some version of the biological species concept: "species are groups of interbreeding natural populations that are reproductively isolated from other such groups" (Mayr 1970). Under this view, particular species are seen first and foremost as gene pools. Other important unifying factors such as genealogy, unitary ecological roles, and morphological distinctness are often just assumed to be congruent with groupings defined by some measure of potential interbreeding.

A number of variations on the basic biological species theme exist, which we would argue represent only minor wrinkles. For example, Paterson (1985) placed emphasis on shared mate recognition systems for defining species rather than on isolating mechanisms as in the classic biological species concept. Furthermore, the biological species concept underlies the treatment of species by Ghiselin and Hull (e.g., Ghiselin 1987; Hull 1987), who have argued that species should be treated metaphysically as individuals, rather than as classes, in large part because of the integrative effects of gene flow.

It is because of the hegemony of this view of species that asexual organisms have been seen as a particular problem. Most of the above authors pointed out that the biological species concept is not applicable to strictly asexual groups. Ghiselin (1987) has argued further that species do not exist in such asexual groups, by definition.

Species as Phenetic Groups

An opposing view of species, with a history pre-dating the breeding groups view, is that of species as basic phenetic clusters. That is, species are groups of similar organisms, separated from other such groups by discontinuities. This viewpoint, while ancient, still has many champions today, perhaps especially among botanists (e.g., Cronquist 1978; Levin 1979). A similar viewpoint is some-times referred to as the "morphospecies" concept, when only morphological characters are used. However, in the broadest sense, phenetic species can be recognized using behavioral and ecological characters as well, or even genetic ones. It is important to recognize that "phenetic" does not mean "phenotypic." It means grouping by overall similarity which can be, and often is, done with genotypic data.

The phenetic approach *is* equally applicable to sexual and asexual organisms. However, it would seem that for most purposes in evolutionary biology (as distinct from purely identification purposes), such as for studies in biogeography, ecologi-cal partitioning, or speciation, it is necessary to have species that have historical integrity. Ecological and evolutionary theory concerns lineages of organisms, not ahistorical groups of similar organisms. Phenetic methods will only result in the naming of lineages as species to the extent that phenetic methods, in general, are successful in delimiting lineages. All the well-known problems with using phe-netic methods for phylogenetic reconstruction apply (Farris 1983).

Species as Evolutionary or Ecological Groups

One concept recognizing species as lineages was originally proposed by Simpson (1961), who extended the notion of unitary evolutionary role over geo-logic time to define the evolutionary species concept. Wiley (1978) modified this definition to include evolutionary stasis of lineages and to exclude anagenetic speciation. Van Valen (1976) emphasized the ecological role of species in his ecological species concept. Still implicit in both evolutionary and ecological species concepts, however, is the view of species as breeding groups and the role of gene flow in providing the major underlying cohesive mechanism. In applying evolutionary or ecological concepts to asexual species (and perhaps in most applications to sexual species), recognition of evolutionary or ecological units is based on overall similarity, following the same procedures used in defin-ing species as phenetic groups.

The more recent "cohesion" species concept of Templeton (1989) was an important step forward in the sense that many potential cohesive mechanisms (e.g., gene flow, ecological or developmental constraints) are considered. Templeton's concept is explicitly applied to asexual organisms. One drawback, however, is that Templeton did not explicitly address the issue of how to ensure that species recognized under his concept will be phylogenetic lineages.

Species as Phylogenetic Groups

In contrast to the viewpoints previously discussed, an explicitly phylogenetic view of species is a relatively recent development, dating back to the late 1970s (critical papers include Bremer and Wanntorp 1979; Rosen 1978, 1979). The central issue, i.e., the lack of necessary correspondence between the biological species concept or the phenetic species concept and monophyletic lineages, has only become clear since the rise of phylogenetic systematics (cladistics), and the clear distinc-tion between symplesiomorphy and synapomorphy. Indeed, many still apparently equate biological species with genealogical entities (e.g., Hull 1987).

A phylogenetic species concept has been proposed and elaborated in two different versions, united by the goal of separating phylogenetic and taxonomic patterns from specific evolutionary processes to the extent that this is possible. The two versions differ in detail on how such a separation can be made. Cracraft (1983, 1989), building on the species concepts of Eldredge and Cracraft (1980) and Nelson and Platnick (1981), defined species as: "the smallest diagnosable cluster of individual organisms within which there is a parental pattern of ancestry and descent." By this he meant that the species category should be used to name the smallest lineages recognizable in a given group. He did, however, retain the qualification that a species must also be "a reproductive community," having some degree of reproductive cohesion (Cracraft 1983).

Mishler and Donoghue (1982; Donoghue 1985; Mishler 1985) and Mishler and Brandon (1987) developed a similar, but more elaborate phylogenetic species concept: "a species is the least inclusive taxon recognized in a classification, into which organisms are grouped because of evidence of monophyly (usually, but not restricted to, the presence of synapomorphies), that is ranked as a species because it is the smallest 'important' lineage deemed worthy of formal recognition, where 'important' refers to the action of those processes that are dominant in producing and maintaining lineages in a particular case" (Mishler and Brandon 1987). The most significant difference between this concept and that of Cracraft (and the related "cladistic species concept" of Ridley 1989), is the explicit recognition that two components are needed for a logically complete species concept. The grouping component (monophyly, or the special case of metaphyly as defined and discussed in Mishler and Brandon 1987) is monistic and, therefore, provides a measure of comparability among species from different groups. The ranking component is necessary because lineages exist at all levels of inclusiveness, particularly in groups with limited gene flow (with asexual groups being the extreme), such that the "smallest" diagnosable monophyletic groups may be far smaller than anyone would want to recognize formally (cell lineages within long-lived clones, for example). It is argued that the ranking component must be pluralistic, in that its application in particular cases should depend on the biology of the group of organisms involved, i.e., the particular mix of reproductive, developmental, or ecological processes causing cohesion of species (e.g., see Templeton's cohesion criteria, cited above).

The latter version of the phylogenetic species concept, unlike the former, is more clearly designed to include asexual as well as sexual organisms. Both types of organisms form lineages (in fact, asexual reproduction leads to much cleaner, strictly diverging lineages). The grouping component is the same; the ranking component applied in different cases may well be different, but is free to follow what becomes known about processes acting to maintain distinctive lineages. The important point relevant to the present symposium is that it is quite possible that these processes are actually quite similar across sexual and asexual groups. Prevailing expectations about the species situation in sexual vs. asexual groups may have been based on a false dichotomy.

Debates about proper species concepts are not merely semantic arguments; they are critical for progress in understanding evolutionary biology. The use to which we can rightly put our species depends on what they represent in the real world. We can be sure that a multiplicity of processes is involved in the origin and maintenance of groupings of organisms. Therefore basing a definition of species on some particular process is dangerous; basing a definition on non-genealogical criteria is misleading, at best.

Expected Patterns of Species Distinctness

Throughout this symposium, we use "clonal" and "asexual" broadly to refer to species with predominantly uniparental reproduction (including vegetative reproduction, parthenogenesis, apomixis, or self-fertilization) with obligate asexuality as the extreme. Despite the common argument that any amount of sexual reproduction makes integration of a species by gene flow likely, we see no hard and fast line in expected variation patterns between asexual and sexual organisms. Infrequent sexual reproduction, especially if localized geographically, can produce situations indistinguishable from obligate asexuality (see discussion by Templeton 1989).

Theoretically, asexual species are predicted to exhibit less variability within populations and greater differentiation among populations than predominantly sexual species. Divergence during speciation should be greater in sexual reproducers (e.g., Grant 1981). These predictions arise as a result of the clonal structure of asexual populations, the internal variability of which is limited due to lack of genetic recombination. Somatic mutations would be expected to accumulate within clones, causing greater differentiation among populations. Reduced overall amounts of recombination among populations should result in reduced distinctness of species.

Use of different species concepts will result in different measures of species distinctness and thereby have serious consequences for recognition of the predicted patterns. If a biological concept is assumed, distances between species can be measured using some measure of reproductive compatibility. If a phenetic concept is applied, various statistical measures of phenotypic distance (e.g., Mahalanobis' Distance) can be used, or several different measures of genotypic distance, including electrophoretic patterns or DNA hybridization. Under a phylogenetic concept, distinctness is defined on the basis of the number of autapomorphies.

Processes of Cohesion and Integration of Species

The notion of cohesion or integration of species has had many meanings, depending in part on differing views of evolutionary processes (see exhaustive review by Van Valen 1982; also Templeton 1989). There are three large classes of causal factors that have been advanced to explain cohesion or integration of species. The first and foremost is gene flow. Starting in the 1960s, a second class of causal factors was increasingly recognized, involving stabilizing natural selection (e.g., Ehrlich and Raven 1969). Then more recently, beginning in earnest in the 1970s, a third class of causal factors was recognized as potentially important in explaining coherence of lineages in general, involving epigenetic or developmental constraints (Alberch 1982; Oster and Alberch 1982; Rachootin and Thomson 1981; Raff and Kaufman 1983; Smith et al. 1985). An explicit application of this third class of factors to the species level was made by Eldredge and Gould (1972), Mishler (1985), Templeton (1989), and Van Valen (1982).

One way to simplify the situation is via the distinction made by Mishler and Brandon (1987), who restricted "integration" to refer to those processes causing active interaction among parts of an entity, such as gene flow or density-dependent natural selection. They restricted "cohesion" to refer to those processes causing an entity to behave as a whole with respect to some process, such as developmental canalization or density-independent natural selection. They pointed out that species recognized under their phylogenetic concept may be integrated, cohesive, both, or neither. Furthermore, the integrative or cohesive processes responsible for particular phylogenetic patterns observed will vary from case to case.

As a fourth alternative, it is important to keep in mind that the mere observation of "clustering" in taxonomic space (i.e., the presence of groups of similar organisms, with discontinuities among such groups; Hutchinson 1968) does not necessarily require a causal "external" explanation such as selective constraints. As discussed by Bookstein (1988), simple random walk processes can produce such discontinuous patterns. Therefore, accidents of history may well explain patterns of discontinuities in morphological variation in some cases, without invoking any integrative or cohesive processes.

Devising empirical tests to determine which of these major classes of processes, if any, are operating in any particular case is difficult (more below). However, obligately asexual organisms may be particularly good cases for study, because one important class of processes (gene flow) can be factored out immediately.

Speciation

The prevailing definition of "speciation" in the literature of evolutionary biology involves the origin of reproductive isolation. Controversies in the speciation literature seem merely to involve different possible geographic, ecological, or genetic modes of acquisition of reproductive isolation (e.g., Coyne and Barton 1988; Templeton 1981), rather than questions about the relevance of isolation per se. Based on the considerations raised above, however, it might be argued that focusing on a single process risks missing what is actually going on.

Under the phylogenetic species concept, reproductive isolating mechanisms are recognized as potentially important factors, but as neither necessary nor sufficient for speciation. Speciation is instead the origin of distinctive new lineages, by a breakdown in any of a number of integrative and/or cohesive mechanisms in "parent" lineages. It might involve a key regulatory mutation producing a new morphology or ecology, or the release of selective constraints in some geographically or ecologically marginal population.

This view of speciation allows the process to be equally well defined for both sexual and asexual organisms (see also White 1978). Study of processes of speciation in asexual organisms again may be of special interest; breeding biology can be factored out, allowing the discovery of perhaps more fundamental mechanisms. Paradoxically, the study of asexual speciation might thus demonstrate mechanisms that are also important in sexual groups, mechanisms that have been neglected under the hegemony of the biological species concept.

Suggestions for Future Work

The question of the nature of species in asexual groups has unfortunately been confused with a different question: the adaptive significance of sexual vs. asexual reproduction. Thus, it is often suggested that because asexual reproduction is a poor adaptive strategy for organisms to take, the question of species in asexual organisms is of little importance (e.g., Häuser 1987). There are two things wrong with such an argument. First, it is not at all a settled question whether, or in what cases, sexual reproduction is selectively advantageous (Bell 1982; Bernstein et al. 1981; Glesener and Tilman 1978; Grassle and Shick 1979; Halvorson and Monroy 1985; Smith 1978; Stearns 1987; Thompson 1976). Second, even if it is generally established that sexual reproduction has distinct advantages for organism fitness, it does not follow that species formation in asexual groups is impossible or unimportant. Even a relatively rare case of a wholly asexual, distinct species, even a species that is geologically short-lived because of poor adaptability, provides a phylogenetic pattern that demands explanation. As discussed above, the search for such an explanation might still shed light on general species-level processes, applying even to sexual groups.

What recommendations could be made for empirical studies designed to separate selective from developmental explanations of cohesion in an asexual species? The nature of developmental constraints should be measured through combined morphometric and quantitative genetic studies of organisms grown in constant environments (e.g., Atchley 1987). Constrained morphologic variation in taxonomically important characters in the presence of high genetic variability would suggest that phenotypic variation is decoupled from overall patterns of variation at the genotypic level, and that as a consequence, developmental constraints play a major role in species distinctness. On the other hand, high morphologic variability in taxonomic characters would suggest that developmental constraints were not important. Furthermore, standard experiments estimating fitnesses of given phenotypes in different relevant environments should be designed to quantify selective constraints. Field studies should supplement such controlled experiments, and include descriptions of morphometric variation and genetic population structure as well as habitat preferences and demographic censuses of fecundity and mortality. Such experiments and descriptive field studies would need to focus on populations throughout the geographic range of a species, to assess uniformity of possible cohesive forces.

What properties would make a particular group of organisms an ideal subject for study? Ideally, such studies should be carried out on a monophyletic group of species that is well understood phylogenetically, so that precise statements could be made about sister group relationships and relative ordering of speciation events within the group. The group should be diverse, so that sister group comparisons could be made among species with different combinations of breeding systems, niche widths, and developmental canalization. The discovery of a clade composed of strictly asexual species would facilitate factoring out the role of gene flow. It would be necessary to have a detailed understanding of reproductive biology in nature, to be sure that interbreeding could be eliminated as an integrative factor in particular species. The group should be experimentally tractable: easy to grow under defined conditions, with short generation time, and readily observable in the field.

As difficult as such studies would be to design, they are necessary for a true understanding of species-level biology. Despite the apparent emphasis of evolutionary biologists since Darwin, rather little hard information is available on the origin of species, per se, as distinct from population processes. We have suggested that this is at least partly due to a lack of conceptual clarity about what species are. Asexual groups may be the key to understanding this critical, but neglected part of evolutionary biology.

ACKNOWLEDGMENTS

We thank Joel Cracraft for his help in organizing the symposium. J. Cracraft and S. D. Cairns read the manuscript and provided helpful comments. We are grateful to the Society of Systematic Zoology and the American Society of Zoologists for providing travel funds to symposium participants.

LITERATURE CITED

ALBERCH, P. 1982. Ontogenesis and morphological diversification. Amer. Zool. 20:653–667.

ATCHLEY, W. R. 1987. Developmental quantitative genetics and the evolution of ontogenies. Evolution 41:316–330.

BELL, G. 1982. The masterpiece of nature: The evolution and genetics of sexuality. Berkeley: Univ. California Press.

BERNSTEIN, H., G. S. BYERS, and R. E. MICHOD. 1981. Evolution of sexual reproduction: Importance of DNA repair, complementation, and variation. Amer. Nat. 117:537–549.

BOOKSTEIN, F. L. 1988. Random walk and the biometrics of morphological characters. Evol. Biol. 23:369–398.

BREMER, K. and H. E. WANNTORP. 1979. Geographic populations or biological species in phylogeny reconstruction? Syst. Zool. 28:220–224.

COYNE, J. A. and N. H. BARTON. 1988. What do we know about speciation? Nature 331:485–486.

CRACRAFT, J. 1983. Species concepts and speciation analysis. Curr. Ornith. 1:159–187.

_____. 1989. Speciation and its ontology: The empirical consequences of alternative species concepts for understanding patterns and processes of differentiation. Pp. 28–59 in Speciation and its consequences, eds. D. Otte and J. A. Endler. Sunderland, Massachusetts: Sinauer Associates.

CRONQUIST, A. 1978. Once again, what is a species? Pp. 3–20 in Biosystematics in agriculture, ed. J. A. Romberger. Montclair, New Jersey: Allanheld & Osmun.

DONOGHUE, M. J. 1985. A critique of the biological species concept and recommendations for a phylogenetic alternative. Bryologist 88:172–181.

EHRLICH, P. R. and P. H. RAVEN. 1969. Differentiation of populations. Science 165:1228–1232.

ELDREDGE, N. 1985. Unfinished synthesis. Biological hierarchies and modern evolutionary thought. New York: Oxford Univ. Press.

_____ and J. CRACRAFT. 1980. Phylogenetic patterns and the evolutionary process. New York: Columbia Univ. Press.

_____ and S. J. GOULD. 1972. Punctuated equilibria: An alternative to phyletic gradualism. Pp. 82–115 in Models in paleobiology, ed. T. J. M. Schopf. San Francisco: W. H. Freeman and Co.

FARRIS, J. S. 1983. The logical basis of phylogenetic analysis. Pp. 7–36 in Advances in cladistics, vol. 2, eds. N. I. Platnick and V. A. Funk. New York: Columbia Univ. Press.

FROST, D. R. and J. W. WRIGHT. 1988. The taxonomy of uniparental species, with special reference to parthenogenetic _Cnemidophorus_ (Squamata: Teiidae). Syst. Zool. 37:200–209.

GHISELIN, M. T. 1987. Species concepts, individuality, and objectivity. Biol. Phil. 2:127–143.

GLESENER, R. R. and D. TILMAN. 1978. Sexuality and the components of environmental uncertainty: Clues from geographic parthenogenesis in terrestrial animals. Amer. Nat. 112:659–673.

GRANT, V. 1981. Plant speciation. New York: Columbia Univ. Press.

GRASSLE, J. F. and J. M. SHICK. 1979. Introduction to the symposium: Ecology of asexual reproduction in animals. Amer. Zool. 19:667–668.

HALVORSON, H. O. and A. MONROY, eds. 1985. The origin and evolution of sex. New York: Alan R. Liss.

HÄUSER, C. L. 1987. The debate about the biological species concept – A review. Z. Zool. Syst. Evolut.-forsch. 25:241–257.

HOLMAN, E. W. 1987. Recognizability of sexual and asexual species of rotifers. Syst. Zool. 36:381–386.

HULL, D. L. 1987. Genealogical actors in ecological roles. Biol. Phil. 2:168–184.

HUTCHINSON, G. E. 1968. When are species necessary? Pp. 177–186 in Population biology and evolution, ed. R. C. Lewontin. Syracuse, New York: Syracuse Univ. Press.

LEVIN, D. A. 1979. The nature of plant species. Science 204:381–384.

LIDÉN, M. and B. OXELMAN. 1989. Species – Pattern or process? Taxon 38:228–232.

MAYR, E. 1970. Populations, species, and evolution. Cambridge, Massachusetts: Harvard Univ. Press.

MISHLER, B. D. 1985. The morphological, developmental, and phylogenetic basis of species concepts in bryophytes. Bryologist 88:207–214.

_____ and R. N. BRANDON. 1987. Individuality, pluralism, and the phylogenetic species concept. Biol. Phil. 2:397–414.

_____ and M. J. DONOGHUE. 1982. Species concepts: A case for pluralism. Syst. Zool. 31:491–503.

NELSON, G. J. and N. I. PLATNICK. 1981. Systematics and biogeography: Cladistics and vicanance. New York: Columbia Univ. Press.

OSTER, G. and P. ALBERCH. 1982. Evolution and bifurcation of developmental programs. Evolution 36:444–459.

PATERSON, H. E. H. 1985. The recognition concept of species. Pp. 21–29 in Species and speciation, ed. E. S. Vrba. Pretoria: Transvaal Museum.

RACHOOTIN, S. P. and K. S. THOMSON. 1981. Epigenetics, paleontology, and evolution. Pp. 181–193 in Evolution today, eds. G. G. E. Scudder and J. L. Reveal. Pittsburgh: Carnegie-Mellon Univ. Press.

RAFF, R. A. and T. C. KAUFMAN. 1983. Embryos, genes, and evolution: The developmental-genetic basis of evolutionary change. New York: MacMillan.

RIDLEY, M. 1989. The cladistic solution to the species problem. Biol. Phil. 4:1–16.

ROSEN, D. E. 1978. Vicariant patterns and historical explanation in biogeography. Syst. Zool. 27:159–188.

_____. 1979. Fishes from the upland and intermontane basins of Guatemala: Revisionary studies and comparative geography. Bull. Amer. Mus. Nat. Hist. 162:267–376.

SHAPOSHNIKOV, G. C. 1984. Aphids and a step toward the universal species concept. Evol. Theory 7:1–39.

SIMPSON, G. G. 1961. Principles of animal taxonomy. New York: Columbia Univ. Press.

SMITH, J. M. 1978. The evolution of sex. Cambridge: Cambridge Univ. Press.

_____, R. BURIAN, S. KAUFFMAN, P. ALBERCH, J. CAMPBELL, B. GOODWIN, R. LANDE, D. RAUP, and L. WOLPERT. 1985. Developmental constraints and evolution. Quart. Rev. Biol. 60:265–287.

STEARNS, S. C., ed. 1987. The evolution of sex and its consequences. Basel: Birkhäuser Verlag.

TEMPLETON, A. R. 1981. Mechanisms of speciation – A population genetic approach. Annual Rev. Ecol. Syst. 12:33–48.

_____. 1989. The meaning of species and speciation: A genetic perspective. Pp. 3–27 in Speciation and its consequences, eds. D. Otte and J. A. Endler. Sunderland, Massachusetts: Sinauer Associates.

THOMPSON, V. 1976. Does sex accelerate evolution? Evol. Theory 1:131–156.

VAN VALEN, L. M. 1976. Ecological species, multispecies, and oaks. Taxon 25:233–239.

_____. 1982. Integration of species: Stasis and biogeography. Evol. Theory 6:99–112.

WHITE, M. J. D. 1978. Modes of speciation. San Francisco: W. H. Freeman and Co.

WILEY, E. O. 1978. The evolutionary species concept reconsidered. Syst. Zool. 27:17–26.

3 A Phylogenetic Species Concept

It is important to recognize that the species problem is not a unique one in systematics. More inclusively, there is a *taxon* problem. One first has to decide what taxon names should represent in general, then species-level taxa should be the same kind of things – just the least inclusive. For me, the arguments made for the Hennigian phylogenetic approach to systematics (Nelson 1973, Farris 1983) have always been compelling. Our classification systems are obviously human constructs, meant to serve certain purposes of our own: communication, data storage and retrieval, predictivity, and most importantly capturing units functioning in processes (Mishler 2009). The most important process in biology is evolution, therefore these purposes are best served by a classification system that reflects our best understanding of descent with modification. Thus the field of systematics, in general, has settled on restricting the use of formal taxonomic names to represent phylogenetically natural, monophyletic groups. Following this general spirit in the 1980's and 1990's, my concern was in developing an approach to species that fits into the Hennigian phylogenetic system.

GROUPING VS. RANKING

There are two necessary parts of *any* species definition. The criteria by which organisms are grouped into taxa must be specified, as well as the criteria by which a taxon is ranked as a species rather than some other taxonomic level. Mishler & Donoghue (1982), Mishler & Brandon (1987), and Mishler & Theriot (2000) argued that the *grouping* criterion that should be used for species is monophyly (there can be special difficulties applying the concept of monophyly at this level, of course; see discussion below). Under this view, synapomorphies are considered to be the necessary empirical evidence for grouping phylogenetic species, as for phylogenetic taxa at all levels.

We argued that the *ranking* decision should include consideration of processes acting in the particular group in question. The question of what causes integration/cohesion of lineages has many possible answers: breeding relationships and patterns of gene flow are relevant here, but so are ecological niches, stabilizing selection, and developmental constraints. The decision about applying a species rank could also involve practical criteria such as the amount of character support for a group, the origin of a distinctive mating system at a particular node, or the acquisition of exclusivity (a condition in which each allele in a lineage is more closely related to another allele in the lineage than it is to an allele *outside* the lineage; Graybeal 1995). We argued that the ranking decision was pluralistic in that different sets of criteria would be necessary for use in different cases.

MONOPHYLY

Over the history of Hennigian phylogenetic systematics, there have been two distinctly different (yet often confused) perspectives when defining monophyly. Monophyly has often been defined in a *diachronic* sense (i.e., "an ancestor and all of its descendants"). Or, it has almost equally often been defined in a *synchronic* sense (i.e., "all and only descendants of a common ancestor"). The salient distinction is that in the latter the ancestor is not included in the group (it is only used as a referent), while in the former the ancestor is included in the group. The latter approach is better in several ways (Mishler 2010, reprinted below) and is especially necessary for my purposes in discussing species, in order to distinguish the extant group of relatives from their ancestor. Thus the concept of monophyly used here is synchronic: *a group containing all and only descendants of a common ancestor, existing in any one instantaneous slice in time.*

The concept of monophyly is in need of further refinement in light of modern genomic data. Horizontal transfer (reticulation) is much more common in nature than realized 20 years ago. Despite being frequently presented as such, reticulation is not just a problem for the species level; clades at all levels can be subject to horizontal transfer. In the modern genomic world, because of the mounting evidence of horizontal gene transfer at all levels (e.g., Beltrán et al. 2002, Degnan & Rosenberg 2009, Mallet et al. 2016), monophyly of a group of organisms can no longer mean monophyly on every gene tree in its genome (as assumed by earlier generations of cladists including me, before there were data to the contrary). We would have few to no monophyletic groups, at any level, in that strict sense.

Rather, we need a revised concept that I'll call "modern monophyly," which refers to an emergent, ensemble characteristic of organismal descent as discussed by Baum (2007, 2009). Just like a person can get a skin graft or liver transplant from someone else and still be the same person, a biological lineage can pick up genes horizontally and still be the same lineage. The key issue is how much and how often this can happen and still be reasonable to call it the same lineage. Consider two leaky hoses laying next to each other in the yard. Some water from one may end up getting into the other but there are still clearly two hoses. But if there is a **lot** of leakages (i.e., both are completely broken at some point) it becomes just one combined stream of water. We botanists have long been happy thinking that hybridization can lead to either situation, the first we have called "introgression," the second we have called "hybrid speciation."

The *ontology* of monophyly can remain as stated given in Mishler and Brandon (1987): "A monophyletic taxon is a group that contains all and only descendants of a common ancestor, originating in a single event." "Event" here refers to processes of integration and cohesion uniting the ancestor, and connecting it via lineages to its descendants, as discussed in our paper, reprinted below. The change needed is in the *epistemology* of monophyly: modern monophyly requires congruence among a plurality of gene trees to call something at a more inclusive level monophyletic. So congruence, e.g., as measured by concordance (see Steel & Velasco 2014) is not definitional; it is simply an epistemological tool. Congruence is evidence used to point back only to a putative common ancestor of the clade in question, not further back in time, i.e., it is fine to call a clade resulting from a hybridization event a monophyletic group as long as it was one event. So while there can't be multiple origins of a hybrid "species" as widely assumed (e.g., Ptacek et al. 1994, Sampson & Byrne 2012), just like there can't be multiple origins of a non-hybrid taxon, there can certainly be a monophyletic origin of a hybrid clade.

Note also that many of the discordant gene trees observed among related clades are **not** due to horizontal transfer at all, but rather to incomplete lineage sorting. Plus, there are two other major sources of discordant gene trees among related lineages: convergence due to selection, and erroneous reconstruction of gene trees due to rate heterogeneity, etc. In the first two cases (horizontal transfer and lineage sorting) we are talking about "true" incongruence, i.e., genes with truly different histories than their containing lineages. Whereas the second two cases are due to inaccurate gene trees. These cases are conceptually easy to distinguish but very difficult to tell apart in practice. One usually sees **many** incongruent gene trees in any given analysis, but probably a small minority of those are due to horizontal transfer. The bacteriologists like Doolittle (1999) who see horizontal transfer everywhere don't take these other possibilities into account and are thus likely overemphasizing reticulation. It is likely that erroneous reconstruction of gene trees is the basis for most of the incongruence, given problems with orthology assessment, alignment issues, saturation in fast-evolving genes, etc.

Thus monophyly, in this modern sense, refers to the preponderance of gene lineages making up a clade (using the clade-lineage distinction given below). Gene lineages that don't match the pattern of descent shown by the majority of lineages need a different explanation (e.g., horizontal transfer, incomplete lineage sorting, mistaken reconstruction, or convergence) than the majority. Note that this is analogous to the distinction people have long made between homology and homoplasy. In fact, all these sources of incongruence, including horizontal gene transfer, are best viewed as types of homoplasy. It has become clear that even a large amount of incongruence does not preclude cladistic reconstructions of phylogeny.

CLADE VS. LINEAGE

Both of the previous perspectives on defining monophyly have useful applications, if we recognize that the things being defined under the two concepts are fundamentally different things. The diachronic sense defines a *lineage,* while the synchronic sense defines a *clade* (see fig. 4.2). These are not at all the same thing, although their meanings intertwine. A "lineage" is a diachronic concept, a relationship through time among a series of ancestor-descendant replicators. On the other hand, a "clade" is a synchronic concept, an instantaneous snapshot of a lineage cross-section. Monophyly as discussed above only applies to clades.

These two conflicting ways of viewing taxa have translated to considerable controversy in the debates over defining species in a phylogenetic sense. Some concepts view species as synchronic things (clades; Mishler & Donoghue 1982, Mishler & Brandon 1987, Mishler & Theriot 2000) and some view species as diachronic things (lineages; Wiley 1978, de Queiroz 1999). Viewing species as clades avoids several problems faced by the alternative where they are viewed as lineages. Operational criteria have been worked out over the last several decades to test hypotheses about clades using apomorphic characters, whereas no empirical criteria have been proposed to test hypotheses about lineages. In addition, conceptually defining species as clades allows them to fit logically in a hierarchical phylogenetic system where taxa at all more inclusive levels are definitely considered clades.

To summarize, the two papers reprinted below (Mishler & Brandon 1987, Mishler & Theriot 2000), assumed that the current codes of nomenclature employing ranks were to continue, and defined a phylogenetic species concept. First, organisms should be grouped into species on the basis of evidence for monophyly (defined synchronically), as at all taxonomic levels. Second, ranking criteria used to assign species rank to certain monophyletic groups must vary among different organisms, but might well include ecological criteria or presence of breeding barriers in particular cases.

LITERATURE CITED

Baum D. 2007. Concordance trees, concordance factors, and the exploration of reticulate genealogy. *Taxon* 56:417–26.

Baum, D.A. 2009. Species as ranked taxa. *Systematic Biology* 58(1): 74–86.

Beltrán, M., C.D. Jiggins, V. Bull, M. Linares, J. Mallet, W.O. McMillan and E. Bermingham 2002. Phylogenetic discordance at the species boundary: comparative gene genealogies among rapidly radiating *Heliconius* butterflies. *Molecular Biology and Evolution* 19: 2176–2190.

Degnan, J. and N. Rosenberg. 2009. Gene tree discordance, phylogenetic inference and the multispecies coalescent. *Trends in Ecology & Evolution* 24(6): 332–340.

de Queiroz, K. 1999. The general lineage concept of species and the defining properties of the species category. Pp. 49–88 in: *Species, New Interdisciplinary Essays*, R. A. Wilson (ed.). Bradford/MIT Press.

Doolittle W.F. 1999. Phylogenetic classification and the universal tree. *Science* 284: 2124–2129.

Farris, J. S. 1983. The logical basis of phylogenetic analysis. Pp. 7–36 in: *Advances in Cladistics. Vol. 2*, Platnick, N. & Funk, V. (eds.). Columbia University Press, New York.

Graybeal, A. 1995. Naming species. *Systematic Biology* 44: 237–250.

Mallet, J., N. Besansky and M.W. Hahn 2016. How reticulated are species? *BioEssays* 38(2): 140–149

Mishler, B.D. 2009. Three centuries of paradigm changes in biological classification: Is the end in sight? *Taxon* 58: 61–67.

Mishler, B.D. 2010. Species are not uniquely real biological entities. Pp. 110–122 in: *Contemporary Debates in Philosophy of Biology*, F. Ayala and R. Arp (eds.). Wiley-Blackwell, Weinheim, Germany.

Mishler, B.D. and R.N. Brandon. 1987. Individuality, pluralism, and the phylogenetic species concept. *Biology and Philosophy* 2: 397–414.

Mishler, B.D. and M.J. Donoghue. 1982. Species concepts: a case for pluralism. *Systematic Zoology* 31: 491–503.

Mishler, B.D. and E. Theriot. 2000. The phylogenetic species concept sensu Mishler and Theriot: monophyly, apomorphy, and phylogenetic species concepts. Pp. 44–54 In: *Species Concepts and Phylogenetic Theory: A Debate*, Q.D. Wheeler & R. Meier (eds.). Columbia University Press, New York.

Nelson, G. 1973. Classification as an expression of phylogenetic relationships. *Systematic Zoology* 22: 344–359.

Ptacek, M.B., H.C. Gerhardt, and R.D. Sage 1994. Speciation by polyploidy in treefrogs: multiple origins of the tetraploid, *Hyla versicolor Evolution* 48: 898–908

Sampson, A.F. and M, Byrne 2012, Genetic diversity and multiple origins of polyploid *Atriplex nummularia* Lindl. (Chenopodiaceae). *Biological Journal of the Linnean Society*, 105: 218–230

Steel, M. and J.D. Velasco 2014. Axiomatic opportunities and obstacles for inferring a species tree from gene trees. *Systematic Biology* 63: 772–778.

Wiley, E. O. 1978. The evolutionary species concept reconsidered. *Systematic Zoology* 27:17–26.

Individuality, Pluralism, and the Phylogenetic Species Concept[9]

Abstract

The concept of individuality as applied to species, an important advance in the philosophy of evolutionary biology, is nevertheless in need of refinement. Four important subparts of this concept must be recognized: spatial boundaries, temporal boundaries, integration, and cohesion. Not all species necessarily meet all of these. Two very different types of "pluralism" have been advocated with respect to species, only one of which is satisfactory. An often unrecognized distinction between "grouping" and "ranking" components of any species concept is necessary. A phylogenetic species concept is advocated that uses a (monistic) grouping criterion of monophyly in a cladistic sense, and a (pluralistic) ranking criterion based on those causal processes that are most important in producing and maintaining lineages in a particular case. Such causal processes can include actual interbreeding, selective constraints, and developmental canalization. The widespread use of the "biological species concept" is flawed for two reasons: because of a failure to distinguish grouping from ranking criteria and because of an unwarranted emphasis on the importance of interbreeding as a universal causal factor controlling evolutionary diversification. The potential to interbreed is not in itself a process; it is instead a result of a diversity of processes which result in shared selective environments and common developmental programs. These types of processes act in both sexual and asexual organisms, thus the phylogenetic species concept can reflect an underlying unity that the biological species concept can not.

Keywords

Species concepts, individuality, pluralism, monophyly, cladistics, phylogeny.

Introduction

The species question continues to be of central interest to biologists and philosophers. Perhaps surprisingly for a topic that has been discussed so frequently for so long, new insights and original interpretations continue to emerge. In our opinion, however, widespread confusion remains on several important points.

[9] B.D. Mishler and R.N. Brandon. 1987. Individuality, pluralism, and the phylogenetic species concept. Biology and Philosophy 2: 397–414. [reprinted by permission]

Our purpose here cannot be to provide a general review of the subject (for which see Mayr, 1982, 1987). We wish instead to concentrate on the flurry of recent philosophically-oriented papers on species (Bernier, 1984; Ghiselin, 1987; Haffer, 1986; Holsinger, 1984, in press; Hull, 1984, 1987; Kitcher, 1984a, 1984b; Kitts, 1983, 1984; Mayr, 1987; Rieppel, 1986; Ruse, 1987; Sober, 1984; Williams, 1985), and to make several points. First, the distinction between individuals and classes is an oversimplification; at least four important subparts of the concept of individuality can be recognized. Second, a phylogenetic species concept has recently been elaborated that can simultaneously and rigorously meet the needs of systematists and evolutionary biologists. Species delimited in this way will never be classes, yet they will often not be fully individuals either. Third, in order to apply this concept, the usually unrecognized distinction between "grouping" and "ranking" components of a species concept must be realized and the appropriate meanings of "pluralism" and "monophyly" with respect to species must be appreciated.

Individuality

The "radical solution to the species problem" advocated by Ghiselin (1974) and Hull (1976) was to consider species as individuals rather than as classes. By "individuals" they meant entities that are "spatiotemporally localized, well-organized, cohesive at any one time, and continuous through time" (Hull, 1987). This idea has been enormously productive as a source of new insights into the species problem. Nevertheless, it is time to move beyond the simple class-individual distinction to a more detailed consideration of properties held by biological entities.[10]

A number of authors have suggested that the class-individual distinction advocated by Ghiselin and Hull is oversimplified and have suggested other ontological categories (Wiley, 1980; Mayr, 1987). Indeed, Hull (1976) himself suggested that a species may fall into some hybrid category that is neither an individual nor a class; but, he claimed, it is at least clear that species are not classes. The last conclusion we find ourselves in complete agreement with. It has been established beyond a doubt, in our opinion, that neither species nor other biological taxa can productively be viewed as sets or classes defined by possession of certain features. We believe that it is possible to define classes that are coextensive with particular biological species (see attempts by Kitcher, 1984a). But such definitions do not add anything to the theoretical insights that have been gained by the "species as individual" concept.

A refinement that can lead to further theoretical insights is to unpack the concept of individuality into important subparts. With regard to evolutionary

---

[10]We should note at the outset that, contrary to the impression one is likely to get from the literature on species-as-individuals, the class-individual distinction is not a distinction taken directly from logic. First, Hull and Ghiselin are using a restricted notion of classes. Something counts as a class for them only if its membership can be specified in a spatiotemporally unrestricted way. Logic places no such restriction on classes. Although Hull (1978) is reasonably clear on this point, not everyone else has been and this has lead to some confusion. Second, the operative notion of "individual" comes more from common sense zoology than from logic.

biology, at least four major sub-concepts of individuality can be recognized. We are not concerned with what sub-concept (or combination thereof) should be called true individuality. Rather we will argue that various kinds of biological entities (including those called species by systematists) will meet various combinations of these criteria of individuality and that it is necessary to distinguish among them. Our concern is to argue against the largely tacit assumption that entities meeting some of these criteria will meet them all.

We have suggested names for these sub-concepts, based on terms that have been used in the literature; other terminologies are clearly possible. It is important to note that the first two of these sub-concepts are different in kind from the second two. The former refers to "patterns," i.e., effects of biological processes, and the latter refers directly to the action of processes. We particularly use species taxa as currently defined for examples here, but will defer our recommendations for proper application of these ideas to species until a later section.

Spatial Boundaries

One important aspect of individuality is the spatial localization of a particular entity. The traditional view of a class is that its members may be present anywhere in the universe, if the proper defining features are present. All known evolutionary processes, however, certainly produce entities at all taxonomic levels that are spatially restricted. Thus it would seem that species taxa, properly named, would always meet this criterion.

Temporal Boundaries

A second important aspect of individuality involves temporal restriction of an entity. A taxon must have a single beginning and potentially have a single end in order to count as an individual under this criterion. Thus, such taxa may not re-originate, even if the second-arising entity is indistinguishable from the first. It should be clear that this criterion can be decoupled from the first. Depending on one's definition of species, taxa could easily be recognized that are spatially, but not temporally, restricted. One example would be repeated polyploid speciation in plants via hybridization (Holsinger,1987). The currently controversial systematic concept of monophyly is relevant here, but we defer discussion until a later section.

Integration

Two very different types of causal interaction between processes and biological entities have been lumped under the concept of individuality, thereby causing confusion. We will argue that these types of causal interactions can and often are disconnected from each other and/or from the resulting patterns discussed above, thus careful distinctions must be made.

We have designated "integration" to refer to active interaction among parts of an entity. In other words, does the presence or activity of one part of an entity matter to another part? Examples of this type of causal interaction include the effect of the heartbeat on the circulatory system of an animal, mating relationships and gene flow within populations and species, and processes of frequency-dependent and density-dependent natural selection. It has been argued

by a number of authors (summarized by Mishler and Donoghue, 1982) that species taxa as currently delimited often do not meet this criterion of individuality
(even though they may meet one or both of the two criteria listed above).

Cohesion

We have designated "cohesion" to refer to situations where an entity behaves as
a whole with respect to some process. In such a situation, the presence or activity of one part of an entity need not directly affect another, yet all parts of the
entity respond uniformly to some specific process (although details of the actual
response in different parts of the entity may be different because of the opera-
tion of other processes). Examples of this type of causal interaction include the
failure of a corporation due to a stock market crash, developmental canalization
in biological systems, and processes of density-independent natural selection.
Clearly, species taxa as currently delimited may show cohesion as defined in this
way, or integration, or both, or neither.

Problems with Application of Individuality to Species

It should be clear from the above examples that despite its philosophical appeal,
the "species as individual" concept developed by Ghiselin and Hull cannot be
applied in its simplistic form to most species taxa as currently delimited, nor, we
would argue, could taxonomic practice be revamped so as to make it generally
applicable (see Mishler and Donoghue, 1982, for further arguments and examples). The major reasons for this inapplicability are two: the plethora of causal
processes acting on biological entities and the lack of correspondence between
either these processes or patterns resulting from them.

As pointed out by Van Valen (1982) and Holsinger (1984) among others, a
great number of processes impinge on organisms and groups of organisms. A
non-exhaustive list would include breeding relationships, competition, geological change, developmental canalization, symbioses, and predation. Entities can
simultaneously behave as individuals with respect to different processes, at different levels of inclusiveness (Holsinger, 1984). Furthermore, groups of organisms defined by aspects of individuality with respect to one process are often
not congruent with groups defined with respect to a second process (Mishler
and Donoghue, 1982).

Mary Williams' recent attempt (1985) to link her concept of "Darwinian subclan" with Ghiselin and Hull's formulation of species as individuals fails for both
of these reasons. Her whole argument rests on the assumption that all biological
species are in the domain of a legitimate interpretation of "Darwinian subclan,"
or in other words, that species are Darwinian subclans. However, this amounts
to the assumption that species are cohesive units with respect to (at least some)
selective forces, i.e., that organisms within a species are all acted upon by those
same forces. This flies in the face of much of what is known about selection.
For example, a species ranging over a geographical cline would hardly qualify as
a Darwinian subclan. For a more theoretical example, consider the intrademic
models of kin and group selection (Wilson, 1980). Here the population units that
are cohesive with respect to selection are generally much smaller than the local

population, much less the entire species. It is possible, even likely, that species will
be Darwinian subclans for some period of their existence (especially at their origin),
but this does not help Williams' argument. She needs this to be generally true.
However, current knowledge of evolutionary processes does not back her up.

The upshot is that species taxa often are not integrated or cohesive because
of particular selective regimes. Other processes causing integration and/or
cohesion of species taxa include gene flow and developmental canalization
(Van Valen, 1982; Mishler, 1985). As mentioned above, species taxa as currently recognized may not be integrated or cohesive in any sense (although as
will be discussed below, this situation might be changed by revision of taxonomic practice). Furthermore, there is no reason to believe that reproductive
processes and selective processes pick out the same units in nature (Mishler
and Donoghue, 1982; Holsinger, 1984) – a correspondence necessary to relate
Williams' Darwinian subclans to Mayr's biological species concept.

To summarize this section, it is useful to consider the nature of various examples of biological entities with differing degrees and aspects of individuality, to
drive home the point that application of the simple dichotomy between individuals and classes has obscured important distinctions. Are there important
biological groupings that are spatiotemporally localized but neither integrated
nor cohesive? Yes, monophyletic higher taxa, called historical entities by Wiley
(1980), and Darwinian clans as formalized by Williams (1970), would usually fit
such a description. Mayr (1987) suggests that species often represent an intermediate kind of entity (which he terms a "population") that have spatiotemporal localization but weak integration and cohesion. Thus the distinction made
above can admit to differing degrees of integration or cohesion, ranging from
strong (in a paradigmatic individual organism) to weak or absent.

Are there important biological groupings that are integrated and/or cohesive but not spatiotemporally localized? Yes, groups defined by their participa-
tion in processes, such as plant communities, pollinator guilds, trophic levels,
mixed-species feeding flocks, or C_4 photosynthesizers, may be highly integrated,
cohesive, or both, and yet lack any temporal boundaries. Further examples are
given by polyphyletic or paraphyletic taxonomic groupings. Such groups may
be cohesive because of ecological factors or shared developmental programs,
but lack a unique beginning (in the case of polyphyletic groups) or a unique
end (in the case of both kinds of groups). Integration and cohesion do seem
to require some form of spatiotemporal connectedness, but, as our examples
illustrate, this does not imply temporal boundaries. Does it strictly imply spatial
boundaries? We think it does; in any case, we cannot think of any plausible
examples of integrated and/or cohesive entities lacking spatial boundaries.

The Phylogenetic Species Concept

The search for a satisfactory concept of species is complicated by the need to
simultaneously reconcile recent advances in evolutionary theory, with recent
advances in systematic theory, with empirical requirements of objectivity and
testability, and with constraints imposed by the formal Linnaean nomenclatorial

system. Before discussing one recently proposed solution, there is a need to introduce and clarify two important subjects: pluralism and the distinction between grouping and ranking.

Pluralism

As a number of authors have pointed out, controversies in evolutionary biology over causal agents generally do not involve claims that all but one favored agent are impossible. Rather, a number of causal agents are acknowledged to be possible and controversy centers around which agent is the "most important" (Gould and Lewontin, 1979; Beatty, 1985).

The result of this situation in evolutionary biology has been a number of calls for "pluralism," meaning generally to keep an open mind about which particular causal agent is to be invoked as an organizing principle in any particular case. The case of species concepts has heard similar calls (Mishler and Donoghue, 1982; Kitcher, 1984a, b).

However, in the case of species, two very different sorts of "pluralism" have been advocated, thus confusion has resulted. Both sorts of pluralism are based on the fact that many different (and non-overlapping) groups of organisms are functioning in important biological processes (see discussion by Holsinger, 1984,1987). Both sorts of pluralism deny that a universal species concept exists. However, they differ in their application to particular biological cases. Kitcher's (1984a, b) brand of pluralism implies that there are many possible and permissible species classifications for a given situation (say the *Drosophila melanogaster* complex), depending on the needs and interest of particular systematists. In contrast, Mishler and Donoghue's (1982) brand of pluralism implies that a single, optimal general-purpose classification exists for each particular situation, but that the criteria applied in each situation may well be different. This latter meaning of pluralism, we would argue, is close to the use of the term by Gould and Lewontin (1979). Furthermore, we would also argue that its use results in perfectly reasonable and rigorous scientific solutions to particular problems. The only caveat is that problems (such as difficult species complexes) that seem at least superficially similar may require different criteria for solution.

Ghiselin (1987) has unfortunately confused these two uses of "pluralism" and tarred them both with a broad brush. Also, unfortunately, he has engaged in ad hominem attacks (by suggesting that pluralists are lazy, incompetent, dishonest, and generally not engaged in science at all) and fallacious arguments. Despite his unsupported assertion that the biological species definition is "fully applicable to plants," numerous botanists (and others) have published careful empirical and theoretical analyses of the difficulties with applying the biological species concept (see Mishler and Donoghue, 1982 for references). Problems having to do with lack of correspondence between patterns resulting from different causal processes, and the gradual nature of breeding discontinuities in plants, cannot be waved aside casually.

To further distinguish between the two meanings of "pluralism" and to clarify the proper usage of the term with respect to biological theories, it is necessary to examine connections with the concept of parsimony. It is natural and correct for scientists to have a bias towards monism, because of the fundamental scientific tenet of economy in hypotheses. Hull's (1987) arguments for consistency in using cessation of gene flow as a uniform definition of the species category carry a lot of weight (see also arguments by Sober, 1984). The burden of proof rests squarely on someone who argues that the current domain of explanation of a monistic theoretical concept must be broken into smaller domains, each with its own explanatory concept. Note that this sort of pluralism (which is the sort advocated by Gould and Lewontin, 1979; Mishler and Donoghue, 1982) is "pluralistic" only during the transition as a prevailing monistic concept is broken up. Once controversy settles and the transition is complete, you are left with a greater number of explanatory concepts, each quite monistic within its proper domain. Parsimony considerations weigh in balance against the need to provide proper explanations for biological diversity. As scientists, we strongly attempt to minimize the number of theoretical concepts (to one if possible) allowed to delimit (for example) basic taxonomic units. Yet we should grudgingly grant status to additional concepts if the need for them is proven in particular cases.

This use of pluralism is clearly not the use advocated by Kitcher (1984a, b). He implies a sort of "permanent pluralism," where an indefinitely large number of theoretical concepts (limited only by interests of particular biologists) remain acceptable within a single domain. We share the skepticism of Sober (1984), Hull (1987), and Ghiselin (1987) towards this meaning of pluralism. Its use with respect to species concepts would seem to rob systematics of any objective way of choosing between conflicting classifications or of any use of species as units of comparison. Therefore, in what follows we use "pluralism" in the sense of Mishler and Donoghue (1982).

Grouping Versus Ranking

All species concepts must have two components: one to provide criteria for placing organisms together into a taxon ("grouping") and another to decide the cut-off point at which the taxon is designated a species ("ranking"). This distinction (as detailed by Mishler and Donoghue, 1982; Donoghue, 1985; Mishler, 1985), has often not been recognized (but see a similar distinction made by Mayr, 1982:254). Taking the biological species concept as an example, its grouping component is "organisms that interbreed." But since such groups are found at many levels of inclusiveness, especially if "potentially interbreeding" is added to the grouping criterion, a ranking component is needed which usually is something like "the largest grouping in which effective interbreeding occurs in nature."[11]

Since both components are implicit in any adequate species concept, confusion is likely to result if the distinction between them is ignored. Thus Hull's (1987) argument that using patterns of gene flow to define species will result in "a consistently genealogical perspective" is unsound. It depends on whether

---

[11] As pointed out by Hull (pers. comm.), when the distinction between grouping and ranking has previously been made, it was often blurred. This may often be because researchers use variations on the same theme for both grouping and ranking; e.g., patterns of morphological similarity or of gene exchange. As will be apparent below, we advocate distinctly different criteria for grouping than for ranking.

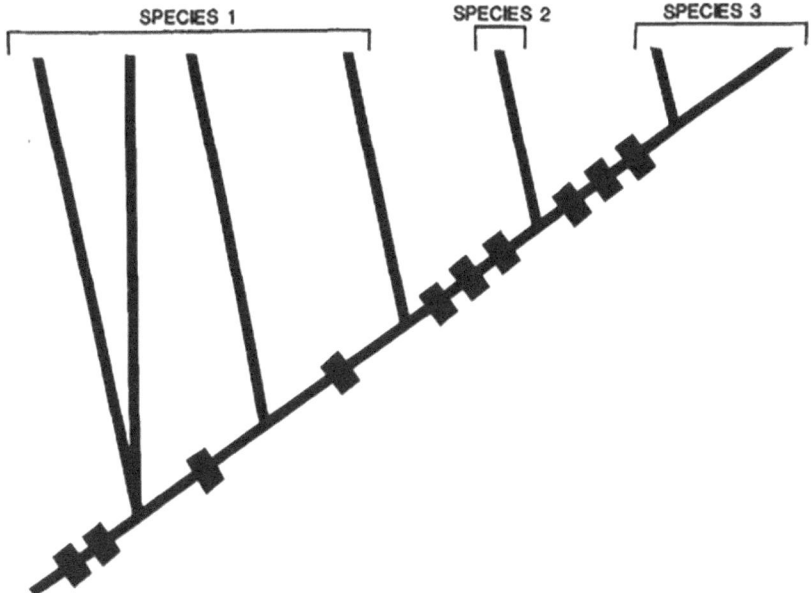

FIGURE 3.1 A hypothetical cladogram showing three named species. Synapomorphies are shown as cross-bars; autapomorphies are not shown. Species 1 is paraphyletic.

reproductive criteria are used for grouping or for ranking. Both Rosen (1979) and Donoghue (1985), among others, have nicely shown that the use of reproductive criteria in grouping can easily result in non-monophyletic taxa, in contrast to the genealogical units Hull (along with us) hopes for. The "recognition concept of species" (Paterson, 1985), wherein species are defined by the possession of a common fertilization system, suffers from a similar problem in that non-monophyletic taxa often result (see Fig. 3.1, where Species 1 may well be definable by reproductive criteria but is not monophyletic).

Further objections to various prevailing species concepts have been given by Mishler and Donoghue (1982), Donoghue (1985), and Mishler (1985). These authors made the following points: (1) None of the dozens of species concepts held currently by various authors can provide grouping criteria able to produce truly genealogical species classifications (including, curiously enough, species concepts advocated by cladists, a group dedicated to genealogical classification). (2) In order to reflect the diversity of causal agents directing evolutionary differentiation in different lineages, no universal ranking criterion can be found.

An Alternative Concept of Species

An alternative perspective on species as genealogical, theoretically significant taxa has been developed by Mishler and Donoghue (1982), Donoghue (1985), and Mishler (1985) and called the "phylogenetic species concept" (not to be confused with the concept proposed by Cracraft, 1983, with the same name). This concept explicitly recognizes a grouping and a ranking component, is monistic with respect to grouping yet pluralistic (in the sense advocated above) with respect to ranking, and produces species taxa with at least some aspects of individuality.

The grouping criterion advocated by Mishler and Donoghue is monophyly in the cladistic sense. Further discussion of the meaning of "monophyly" is needed (see below), because the term is not normally applied to species in a substantive way by cladists. For now, it suffices to say that "monophyly" here is taken to refer to a grouping that had a single origin and contains (as far as can be empirically determined) all descendants of that origin.

Monophyletic groupings as roughly defined above exist at all levels of inclusiveness, thus a ranking criterion for species is needed as the basal systematic taxon (i.e., the least inclusive monophyletic group recognized in a particular classification). It is here that Mishler and Donoghue have advocated a pluralistic adjustment in the number of ranking criteria allowable for consideration in particular cases. They argued that the currently favored monistic ranking concept of absolute reproductive isolation is not the most appropriate for all groups of organisms. The ranking concept to be used in each case should be based on the causal agent judged to be most important in producing and maintaining distinct lineages in the group in question. The presence of breeding barriers might be used, but so might selective constraints or the action of strong developmental canalization (Mishler, 1985). In the great majority of cases, little to nothing is actually known about any of these biological aspects. In such cases grouping (estimation of monophyletic groups) will proceed solely by study of patterns of synapomorphy (i.e., shared, derived characters), and a practical ranking concept must be used until something becomes known about biology. This preliminary and pragmatic ranking concept will usually be the size of morphological gaps (i.e., number of synapomorphies along any particular internode of a cladogram) in most cases, a concept in accord with current taxonomic practice.

The phylogenetic species concept (PSC) of Mishler and Donoghue can be summarized as follows:

> A species is the least inclusive taxon recognized in a classification, into which organisms are grouped because of evidence of monophyly (usually, but not restricted to, the presence of synapomorphies), that is ranked as a species because it is the smallest "important" lineage deemed worthy of formal recognition, where "important" refers to the action of those processes that are dominant in producing and maintaining lineages in a particular case.

Relating the PSC back to the earlier discussion of individuality, it is clear that species so defined (as with monophyletic taxa at all levels) will at least meet the restricted spatiotemporal criterion of individuality. They may or may not be integrated or cohesive. However, these criteria may often prove useful in ranking decisions. Since the strength of integrative or cohesive bonds tends to gradually weaken as more and more inclusive groups of organisms are taken (see for example discussion by Mayr, 1987), it may be possible in many cases to objectively fix the species level as the most inclusive monophyletic group that is integrated or cohesive with respect to "important" processes. Again, "important" has a context-dependent meaning, and will often not refer to reproductive criteria. It may often be difficult to apply this standard, especially if macroevolutionary processes occur (even rarely) involving groups at high taxonomic levels (Gould, 1980; Jablonski, 1986). If so, integrated and/or cohesive groups may occur at much more inclusive levels than anyone would wish to name as basal taxonomic units.

The problem of (at least partial) non-comparability of species taxa in different groups of organisms is a real one (Sober, 1984; Hull, 1987; Ghiselin, 1987). However, as pointed out by Mishler and Donoghue (1982), this has always been the case, despite the fact that many users of species taxa – ecologists, philosophers, paleobiologists, biogeographers, for example – remain blissfully unaware. This difficult situation has not come about because (as suggested by Ghiselin, 1987) systematists working with organisms other than birds are incompetent, but rather reflects a fact of nature. The pluralistic ranking concept of the PSC was proposed to allow different biological situations to be explicitly treated. Persons interested in studying some biological processes simply cannot avoid the responsibility of learning enough about the systematics of the organisms they are studying to ensure that the entities being compared are truly comparable with respect to that process.

To take one example that has been widely recognized (Mayr, 1987), asexual organisms present insurmountable difficulties for the biological species concept. One proposed solution has been to deny that such organisms form species (Bernstein et al., 1985; Eldredge, 1985; Hull, 1987; Ghiselin, 1987). This reductio ad absurdum of the biological species concept demonstrates how a monistic ranking (and grouping) concept based on interbreeding criteria can obscure actual patterns of diversification. One of us (B.D.M.) happens to work on a genus of mosses (*Tortula*, see Mishler, 1985, for references), in which frequently sexual, rarely sexual, and entirely asexual lineages occur. The interesting thing is that the asexual lineages form species that seem comparable in all important ways with species recognized in the mostly asexual lineages and even in the sexual lineages.[12] It just happens in this case that potential interbreeding or lack thereof seems of little or no importance in the origination and maintenance of diversity. The application of the PSC here is able to reflect an underlying unity that the biological species concept could not.

Indeed there seems to be a fundamental confusion at the heart of the biological species concept and its insistence that only sexual organisms can form species. Potential interbreeding and the lack thereof (i.e., breeding barriers) can be observed in nature and so can be used as a ranking criterion for species. But why should it be so used, or rather, why should it be the only ranking criterion used? We suspect that part of the rationale stems from a confusion over the roles of potential interbreeding and actual interbreeding.

---

[12] A similar result has been arrived at by Holman (pers. comm.) based on comparisons between bdelloid rotifers (which are exclusively parthenogenic) and monogonont rotifers (which occasionally reproduce sexually). Using numbers of synonymous species names as an index of taxonomic distinctness of species, he has shown that bdelloid species are apparently more consistently recognized by taxonomists than are monogonont species.

Actual interbreeding is a process. It results in lineages (but not always lineages important enough to be named species – e.g., short-lived hybrid populations). The process of (actually) interbreeding also inevitably leads to a certain amount of integration. In sexual species, it undoubtedly is one of the important processes holding the species together. But potential interbreeding is not a process and therefore has no effect on the integration or cohesion of species. The dispersed parts of a sexual species are not bound together by this non-process; they may be bound together by sharing common environments or common developmental programs, but they cannot be bound together by "potential interbreeding."

In general, the potential to interbreed is based on organisms sharing common environments and common developmental programs. The processes that result in groups of organisms sharing such features and in discontinuities between such groups are multifarious, and are not restricted to sexual organisms. Organisms share common developmental programs because they share a common ancestor. Reproduction is a relevant process here, but not necessarily sexual reproduction.

It is our argument that the PSC is superior to the biological species concept (or to the evolutionary species concept of Simpson, 1961, and Wiley, 1978, which is similar in these ways to the biological species concept) in two fundamental ways. First, monophyly as a grouping criterion is superior to ability to interbreed because it will lead to a consistently genealogical classification. Second, the pluralistic ranking concept of the PSC is superior to the monistic insistence on breeding barriers of the biological species concept because it can more adequately reflect evolutionary causes of importance in different groups.

Other cladistic species concepts such as the "phylogenetic species concept" of Cracraft (1983) which is very similar to the species concept of Nelson and Platnick (1981) are also inferior to the PSC of Mishler and Donoghue, but for somewhat different reasons. The grouping concept used by the former authors (i.e., a cluster of organisms defined by a unique combination of primitive and derived characters) does not rule out the possibility of paraphyletic species, unlike the PSC (see next section). Furthermore, the concepts of Cracraft and Nelson and Platnick (in addition to the concept of Rosen, 1979, that does use presence of synapomorphies as a grouping criterion) are incomplete in that they lack a ranking criterion. It is not sufficient to say that a species is the smallest diagnosable cluster (Cracraft, 1983) or even monophyletic group, because such groups occur at all levels, even within organisms (e.g., cell lineages). Some judgment of the significance of discontinuities is needed.

Monophyly

One final area in need of clarification is the concept of monophyly. Traditionally, the cladistic definition of monophyly (which we favor) has not been applied to the species level. Hennig (1966) did not do so because he was committed to a biological species concept and thought that there was a clean break at the species level, with reticulating genealogical relationships predominating below and diverging genealogical relationships predominating above. Later cladists (e.g., Wiley, 1981) have followed Hennig and defined a monophyletic taxon as one that originated in a single species and that contains all descendants of that species. Species are taken to be monophyletic a priori, therefore it is argued that they need not possess synapomorphies or really be monophyletic in the sense of higher taxa (e.g., Wiley, 1981). One major reason for this is the supposed problem of "ancestral" species.

It is our view that properly clarified, there are no insurmountable problems with applying the concept of monophyly explicitly to species (as the basal systematic taxon). Furthermore, this application must be carried out in order to have a consistently genealogical classification.

Monophyly should be redefined in such a way as to apply to species:

> A monophyletic taxon is a group that contains all and only descendants of a common ancestor, originating in a single event.

"Ancestor" here refers, not to an ancestral species, but to a single individual. By "individual" here, we do not necessarily mean a single organism, but rather an entity (less inclusive than the species level) with spatiotemporal localization and with either cohesion or integration or both (as defined above). In particular cases, this ancestral individual could be a single organism, a kin group, or a local population. We would argue that it would never be a whole species because we share the widespread view that new species come about only via splitting, not by any amount of anagenetic change.

The originating "event" of a monophyletic group referred to in the definition above could be due to the spatiotemporally restricted action of a number of different causes. These could include, in different cases, the origin of an evolutionary novelty which causes a new monophyletic group to be subject to a different selective regime than the rest of the "parent" species or which causes a disruption of the normal developmental canalization of the "parent" species. These could also include acquisition of an isolating mechanism or even the origin of a new species by hybridization between parts of two "parent" species. This diversity of causes for evolutionary divergence reinforces the need for a pluralistic ranking concept.

Some examples of the application of this concept should clarify the definition. It is thought at the present time that a common mode of speciation is via peripheral isolation. In such a case, the peripherally isolated part of the species, if spatiotemporally localized (say on the same island at the same time) and either cohesive, integrated, or both (say by interbreeding and sharing a common niche), would qualify as a monophyletic group under our definition. This would be true even if several rather unrelated members of the original species were the founders of the peripheral population, as long as the above conditions obtain. On the other hand, if two similar but non spatiotemporally connected peripheral populations (say on two different islands) have been established by members (even closely related ones) of the original species, these two populations would have to be considered as two separate monophyletic groups. They are two separate monophyletic groups because they originated in two different events. Hybrid speciation provides similar examples. If two original species produce a hybrid population in one place (say a single valley) at one time (say in a single breeding season), and if this hybrid population behaves as an integrated and/or cohesive entity, then it is a perfectly good monophyletic group under our definition. However, if similar hybrids are produced elsewhere in the ranges of the two original species, or if hybrids are produced in the same locality but discontinuously in time (i.e., if the first hybrid population goes extinct before the new hybrids are produced), then the separate hybrid populations would have to be considered as separate monophyletic groups and could not be taken together and named as a new species. Note that this conclusion is directly opposite that of Kitcher (1984a:314–315). The implications of our concept of monophyly for the original species in the above examples will be discussed below.

This concept of monophyly is, of course, only a grouping criterion. It does not imply that any particular peripheral isolate or hybrid population must be recognized as a species. It only specifies the genealogical conditions under which such groups can be recognized if the ranking criterion applied in a particular case supports recognition at the species level. The grouping and ranking criteria can thus be seen to interact in producing a species classification. Note that a corollary of the PSC is that not all organisms will belong to a formal Linnaean species since some monophyletic groups (e.g., hybrid populations that arise, but then quickly go extinct) will not be judged to be "important" monophyletic groups. The hybrid organisms in such a case would not formally belong to either original species.

The definition of monophyly given above solves the problem perceived by Hennig (1966), Wiley (1981), and Cracraft (1983) with "ancestral species." No such things exist. Only parts of an original species give rise to new ones, as in the above examples. If a currently recognized species is found to be paraphyletic because parts of it can be demonstrated to be more closely related to another species (Fig. 3.1; see also discussions and diagrams of such a situation in Bremer and Wanntorp, 1979; Avise, 1986), then the paraphyletic species should be broken up into smaller monophyletic species.

Note that if Species 1 (Fig. 3.1) is actually integrated by gene flow, then over time its cladistic structure should approach that of Species 1 in Fig. 3.2. Moreover, over an even longer time in such a truly integrated species, patterns of character distribution should even out such that no autapomorphies remain to distinguish lineages within the species, and Species 1 would be represented in a cladogram by a single line (albeit still without any synapomorphies to distinguish it as a species). In systematic studies, a situation is frequently encountered (Fig. 3.2) in which a number of unresolved lineages exist, one or more of which are deemed worthy of recognition as separate species, and the rest of which have traditionally been considered a species taken together. This type of situation has been confused with paraphyly. However, it is actually a case of a taxon (e.g. Species 1 in Fig. 3.2) with a uncertain status between paraphyly and monophyly. With further study, synapomorphic characters may be found uniting some part of Species 1 with the lineage of Species 2 and 3 (as in Fig. 3.1). If that becomes the case, Species 1 truly is paraphyletic and must be broken up. On the other hand, further study may demonstrate synapomorphies uniting all of the lineages in Species 1, thus making it an unproblematic phylogenetic species.

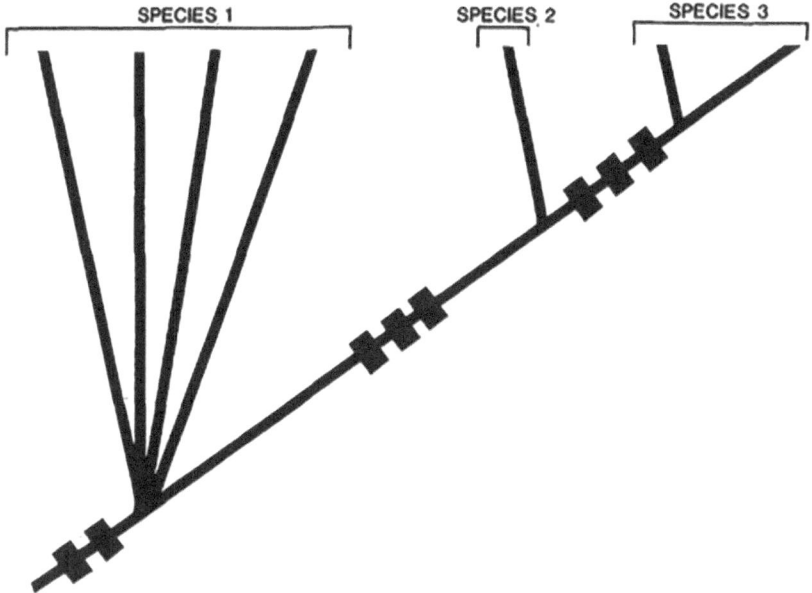

FIGURE 3.2   A hypothetical cladogram showing three named species. Synapomorphies are shown as cross-bars; autapomorphies are not shown. Species 1 is metaphyletic.

It has been cogently argued by Donoghue (1985) that a group such as Species 1 in Fig. 3.2 could acceptably be named a species in a tentative and pragmatic way, pending further study designed to resolve the relationships, as long as a special convention was followed to indicate the uncertain status of the species (Donoghue suggests marking the binomial name of all such species with an asterisk). This solution is practical because it avoids unnecessary naming of highly localized species (if, for example, all recognizable lineages in Species 1, Fig. 3.2, were formally named). It is also probably unavoidable, since if speciation by peripheral isolation occurs frequently, such situations may often be in principle unresolvable, as discussed above. Donoghue (1985) suggested calling this type of species a metaspecies, to clearly distinguish it from a known monophyletic species. Following the prefix he suggested, we suggest the need for a new term, "metaphyly," to refer to the status of groups that are not known to be either paraphyletic or monophyletic. Although beyond the scope of the present paper, this term would clarify similar situations with respect to higher taxa, and may thus prove more widely useful.

Conclusion

The "species problem" as discussed in this paper involves a search for a definition of the basal systematic unit that will be at once practical, provide optimal general-purpose classifications, and reflect the best current knowledge about evolutionary processes. We have claimed that the PSC will fulfill these criteria.

However, we certainly have not claimed that all important biological entities can be recognized using the PSC.

As pointed out clearly by Holsinger (1984), a multitude of interesting biological entities, often non-overlapping, are behaving as (at least partial) individuals with respect to a multitude of interesting processes in any particular group of organisms. While we do need to settle on criteria for recognizing formal taxa for our Linnaean taxonomic system (including species), we are of course in no way prohibited from informally naming and studying other entities of interest that do not fit the formal taxonomic system. That is, as long as different types of entities are explicitly distinguished from each other.

ACKNOWLEDGMENTS

We dedicate this paper to Ernst Mayr, even though he probably disagrees with much of its contents. At different times and in different ways, we both were profoundly affected by our interactions with him during our graduate careers at Harvard. We thank him for his advice, insights, and patience. We also thank David Hull and Marjorie Grene for comments that helped to clarify certain aspects of the paper. Eric Holman kindly allowed us to cite his unpublished data on rotifers.

REFERENCES

Avise, J. C.: 1986, 'Mitochondrial DNA and the Evolutionary Genetics of Higher Animals', Phil. Trans. R. Soc. Lond. B 312, 325–342.

Beatty, J.: 1985, 'Pluralism and Panselectionism', in Asquith, P. D. and Kitcher, P. (eds.), PSA 1984 2, 113–128.

Bernier, R.: 1984, 'The Species as Individual: Facing Essentialism', Syst. Zool. 33, 460–469.

Bernstein, H., Byerly, H. C., Hopf, F. A., and Michod, R. E.: 1985, 'Sex and the Emergence of Species', J. Theor. Biol. 117, 665–690.

Bremer, K. and Wanntorp, H.-E.: 1979, 'Geographical Populations or Biological Species in Phylogeny Reconstruction?', Syst. Zool. 28, 220–224.

Cracraft, J.: 1983, 'Species Concepts and Speciation Analysis', Cur. Ornith. 1, 159–187.

Donoghue, M. J.: 1985, 'A Critique of the Biological Species Concept and Recommendations for a Phylogenetic Alternative', Bryologist 88, 172–181.

Eldredge, N.: 1985, Unfinished Synthesis: Biological Hierarchies and Modern Evolutionary Thought, Oxford University Press, New York.

Ghiselin, M. J.: 1974, 'A Radical Solution to the Species Problem', Syst. Zool. 23, 536–544.

Ghiselin, M. J.: 1987, 'Species Concepts, Individuality, and Objectivity', Biology and Philosophy 2, 127–143.

Gould, S. J.: 1980, 'Is a New and General Theory of Evolution Emerging?' Paleobiology 6, 119–130.

Gould, S. J. and Lewontin, R. C.: 1979, 'The Spandrels of San Marco and the Panglossian Paradigm: A Critique of the Adaptationist Programme', Proc. Roy. Soc. Lond. B 205, 581–598.

Haffer, J.: 1986, 'Superspecies and Species Limits in Vertebrates', Zeit. Zool. Syst. Evolut.-forsch. 24, 169–190.

Hennig, W.: 1966, Phylogenetic Systematics, Univer. Ill. Press, Urbana.

Holsinger, K. E.: 1984, 'The Nature of Biological Species', Phil. Sci. 51, 293–307.
Holsinger, K. E.: 1987, 'Discussion: Pluralism and Species Concepts, or When Must We Agree With Each Other?', Phil. Sci. 54, 480-485.
Hull, D. L.: 1976, 'Are Species Really Individuals?' Syst. Zool. 25, 174–191.
Hull, D. L.: 1978, 'A Matter of Individuality', Phil. Sci. 45, 335–360.
Hull, D. L.: 1984, 'Can Kripke Alone Save Essentialism? A Reply to Kitts', Syst. Zool. 33, 110–112.
Hull, D. L.: 1987, 'Genealogical Actors in Ecological Roles', Biol. Philos. 2, 168–184.
Jablonski, D.: 1986, 'Background and Mass Extinctions: The Alternation of Macroevolutionary Regimes', Science 231, 129–133.
Kitcher, P.: 1984a, 'Species', Phil. Sci. 51, 308–333.
Kitcher, P.: 1984b, 'Against the Monism of the Moment', Phil. Sci. 51, 616–630.
Kitts, D. B.: 1983, 'Can Baptism Alone Save a Species?', Syst. Zool. 32, 27–33.
Kitts, D. B.: 1984, 'The Names of Species: A Reply to Hull', Syst. Zool. 33, 112–115.
Mayr, E.: 1982, The Growth of Biological Thought, Harvard University Press, Cambridge.
Mayr, E.: 1987, 'The Ontological Status of Species', Biology and Philosophy 2, 145–166.
Mishler, B. D.: 1985, 'The Morphological, Developmental, and Phylogenetic Basis of Species Concepts in Bryophytes', Bryologist 88, 207–214.
Mishler, B. D. and Donoghue, M. J.: 1982, 'Species Concepts: A Case For Pluralism', Syst. Zool. 31, 491–503.
Nelson, G. and Platnick, N. I.: 1981, Systematics and Biogeography: Cladistics and Vicariance, Columbia University Press, New York.
Paterson, H. E. H.: 1985, 'The Recognition Concept of Species', in Vrba, E. S. (ed.), Species and Speciation, Transvaal Museum Monograph No. 4, Pretoria, pp. 21–29.
Rieppel, O.:1986, 'Species are Individuals: A Review and Critique of the Argument', Evol. Biol. 20, 283–317.
Rosen, D. E.: 1979, 'Fishes From the Uplands and Intermontane Basins of Guatemala: Revisionary Studies and Comparative Geography', Bull. Amer. Mus. Nat. Hist. 162, 267–376.
Ruse, M.: 1987, 'Species: Natural Kinds, Individuals, or What?', Brit. J. Phil. Sci. 38: 225–242
Simpson, G. G.: 1961, Principles of Animal Taxonomy, Columbia University Press, New York.
Sober, E.: 1984, 'Discussion: Sets, Species, and Evolution: Comments on Philip Kitcher's "Species"', Phil. Sci. 51, 334–341.
Van Valen, L. M.: 1982, 'Integration of Species: Stasis and Biogeography', Evol. Theory. 6, 99–112.
Wiley, E. O.: 1978, 'The Evolutionary Species Concept Reconsidered', Syst. Zool. 27, 17–26.
Wiley, E. O.: 1980, 'Is the Evolutionary Species Fiction? –A Consideration of Classes, Individuals, and Historical Entities', Syst. Zool. 29, 76–80.
Wiley, E. O.: 1981, Phylogenetics: The Theory and Practice of Phylogenetic Systematics, John Wiley, New York.
Williams, M. B.: 1970, 'Deducing the Consequences of Evolution: A Mathematical Model', J. Theor. Biol. 29, 343–385.
Williams, M. B.: 1985, 'Species are Individuals: Theoretical Foundations for the Claim', Phil. Sci. 52, 578–590.
Wilson, D. S.: 1980, The Natural Selection of Populations and Communities, Benjamin/Cummings, Menlo Park.

The Phylogenetic Species Concept (*Sensu* Mishler and Theriot): Monophyly, Apomorphy, and Phylogenetic Species Concepts[13]

Conceptual History

Various attempts have been made at forging a species concept compatible with phylogenetic systematics or cladistics. Several such concepts have been called *the* Phylogenetic Species Concept, thus leading to considerable confusion in the literature. We support one version of the Phylogenetic Species Concept, one that, we will argue, can serve as a synthesis of all versions, but for historical clarity, we will distinguish among different versions, their origins, and motivations (see also discussion by Baum 1992).

Hennig himself apparently held a view on species close to the Biological Species Concept. He defined species as "a complex of spatially distributed reproductive communities" (1966:47). He made an important distinction between tokogenetic relationships (ones that obtain between an "individual and its descendants and predecessors of the first degree" 1966:65) and phylogenetic relationships (ones that obtain between different lineages, "each bounded by two cleavage processes in the sequence of individuals that are connected by tokogenetic relations" 1966:20). In other words, tokogenetic relationships are diachronic, ancestor-descendant connections, whereas phylogenetic relationships are synchronic, sister-group connections.

Hennig's approach, although sound in many respects, errs in our opinion by postulating that there is one single breaking point at which reticulating tokogenetic relationship ends and divergent phylogenetic relationship begins. As we will discuss in detail below, there is not a clear cutoff point at which reticulation

---

[13] B.D. Mishler and E. Theriot. 2000. The phylogenetic species concept sensu Mishler and Theriot: monophyly, apomorphy, and phylogenetic species concepts. In Q.D. Wheeler & R. Meier (eds.), Species Concepts and Phylogenetic Theory: A Debate, pp.44–54. Columbia University Press. [reprinted by permission]

of lineages ceases, and furthermore, the point at which the possibility of reticulation goes to zero is well above the level at which cladistic structure can be reconstructed.

Hennig was also in error, in our opinion, when he used reproductive criteria to group organisms into species. The inappropriateness of using breeding compatibility in cladistic analysis was first pointed out by Rosen (1978, 1979) and Bremer and Wanntorp (1979). The fundamental problem is that the ability to interbreed (potential or actual), as a plesiomorphy by definition, is not a phylogenetically valid grouping criterion. Rosen argued (correctly, we think) that the basis for grouping in a cladistic system should be synapomorphy.

One phylogenetic species concept, based on unique patterns of shared characters, was proposed and defended by Eldredge and Cracraft (1980), Cracraft (1983), and Nixon and Wheeler (1990). For purposes of discussion, in this essay, we shall refer to this general approach as the Phylogenetic Species Concept *sensu* Wheeler and Platnick, abbreviated Phylogenetic Species Concept (W–P). Oddly enough from a phylogenetic standpoint, unlike the concept of Rosen, this concept was explicitly not based on synapomorphy, but rather on a shared combination of characters (tracing back in this way to the species concept of Nelson and Platnick 1981). This approach to phylogenetic species followed the view of Hennig, seeing a fundamental break at the species level between diverging phylogenetic relationships above and reticulating tokogenetic relationships below.

Following up on some of Rosen's suggestions, Mishler and Donoghue (1982) presented and discussed another phylogenetic species concept [but they did not call it such until 1985 (Donoghue 1985; Mishler 1985)]. For purposes of discussion, we shall refer to this general approach as the Phylogenetic Species Concept *sensu* Mishler and Theriot, abbreviated Phylogenetic Species Concept (M–T). We are not particularly concerned about names for concepts; the concepts themselves are the important matter. However, we do not think priority is particularly important in such cases. Although we admit that Cracraft (1983) first used the name Phylogenetic Species Concept, we argue that a name should first and foremost reflect the meaning of a concept, and will point out in the next section that the Phylogenetic Species Concept (M–T) is uniquely suited for a classification based on phylogeny.

The empirical emphasis of Mishler and Donoghue (1982) was twofold. First, they pointed out the obvious noncorrespondence between groupings of organisms defined by different criteria. That is, an ecologically coherent group may be either less or more inclusive than the actively interbreeding group, and neither may correspond to a morphologically and/or genotypically coherent group. Second, as one looks at less inclusive and more inclusive groupings with respect to any one of these factors, there is no fundamental level, no level with some special reality for evolutionary studies. The theoretical emphasis of Mishler and Donoghue (1982) was also twofold. First, organisms should be grouped into species on the basis of evidence for monophyly, as at all taxonomic levels; breeding criteria, in particular, have no business being used for grouping purposes. Second, ranking criteria used to assign species rank to certain monophyletic groups must vary among different organisms, but might well include ecological criteria or the presence of breeding barriers in particular cases.

The two versions of the Phylogenetic Species Concept differ strongly in how they view reticulate relationships and characters in cladistic analysis. The Phylogenetic Species Concept (W–P) argues that units having reticulate relationships are inappropriate for phylogenetic analysis (because that is inherently a study of branching relationships) and that such units can be the unambiguous phylogenetic species. "Fixed" combinations of characters were considered to be the empirical evidence for such units. The Phylogenetic Species Concept (M–T) argues that reticulation can occur throughout the hierarchy of life and so is not a special species problem, but rather one of more general difficulty. Under this view, apomorphies were considered to be the necessary empirical evidence for unambiguous phylogenetic species, as for phylogenetic taxa at all levels.

It might appear from the literature and the above discussion that the two basic versions of the Phylogenetic Species Concept are diametrically opposed. However, the differences can be overemphasized. Looking at the history of ideas and research groups in the manner pioneered by Hull (1988; see also Mishler 1987), it is clear that both in a phylogenetic sense and a phenetic sense, the two Phylogenetic Species Concepts are much closer to each other than either is to phenetic, biological, or evolutionary concepts. Both Phylogenetic Species Concepts have origins in the theory of phylogenetic systematics, and both emphasize that species be diagnosable. Differences in underlying philosophy remain, however. Wheeler and Platnick's Phylogenetic Species Concept has emphasized epistemology in its central focus on character evidence, whereas our Phylogenetic Species Concept has emphasized ontology in its central focus on monophyly. Difficulties in arriving at a synthesis of these two general phylogenetic approaches to species include finding the right balance between primary systematic patterns (i.e., character evidence) and evolutionary process theories. Clearly, it makes no sense to apply a species concept that requires prior specific knowledge of processes (e.g., reproductive behavior or ecological sorting). On the other hand, it is necessary that recognized species taxa be compatible with processes acting to produce phylogenies if phylogenetic classification is to be adopted as the general reference system. A unified Phylogenetic Species Concept can be proposed, based primarily on our Phylogenetic Species Concept in terms of its generalized ontological view about the meaning of phylogenetic criteria at any hierarchical level, but also incorporating the epistemological focus on character evidence from the Phylogenetic Species Concept (W–P).

Definition

The following paragraph provides a formal definition of our Phylogenetic Species Concept (based primarily on that of Mishler and Brandon 1987). The definition is complex, but then again so are the issues involved in producing hierarchical classifications from phylogenies:

> A species is the least inclusive taxon recognized in a formal phylogenetic classification. As with all hierarchical levels of taxa in such a classification, organisms are

grouped into species because of evidence of monophyly. Taxa are ranked as species rather than at some higher level because they are the smallest monophyletic groups deemed worthy of formal recognition, because of the amount of support for their monophyly and/or because of their importance in biological processes operating on the lineage in question.

Some elaboration of terms from this definition is needed (see also Mishler and Brandon 1987). *Monophyly* is defined synchronically, following the "cut method" of Sober (1988), as all and only descendants of a common ancestor existing in any one slice in time. The ancestor is not an ancestral species, but rather a less inclusive entity such as an organism, kin group, or population that had spatiotemporal localization and cohesion/integration (as discussed by Mishler and Brandon 1987). The synchronic approach is necessary to avoid the time paradoxes that arise when classifying ancestors with descendants (see discussion by Hennig 1966). Given that ontology, the evidence required for a hypothesis of monophyly is primarily corroborated patterns of synapomorphy (but may include other factors, such as geography).

The ranking decision (Mishler and Donoghue 1982; Donoghue 1985; Mishler 1985, Mishler and Brandon 1987) can involve practical criteria such as the amount of support for a putative group (e.g., number and "quality" of synapomorphies, bootstrap percentage, or decay index) and may also involve biological criteria in better-known organisms (e.g., the origin of a distinctive mating system at a particular node). This ranking decision is forced because systematists have legislatively constrained themselves to use a ranked Linnaean hierarchy. A larger issue is the recent call for reforming the Linnaean system by De Queiroz and Gauthier (1992) to remove the concept of ranks. Such a move would decrease the arbitrariness of ranking decisions at the species level as well, but the implications of this are beyond the scope of this paper (see Mishler 1999 for a discussion of the implications); we assume here that the current Linnaean system of ranked classifications is to remain in place.

Phylogenetic trees are the primary result of systematic study; they are hypotheses about nature, and thus "real" in that sense. However, any application of fixed names to phylogenetic trees (which result from continuous processes of divergence and reticulation) has to be arbitrary to some extent (particularly ranking). Grouping (based on monophyly) will be less arbitrary, but will still involve ancillary decisions about character homology and about how much support is necessary before one believes a hypothesis of monophyly. The main reason for providing a classification (beyond simply presenting the phylogeny, which could otherwise speak for itself) is to give a convenient handle, a name, for those monophyletic groups that we need to discuss or about which we need to record data. We need to name distinctive lineages as part of the process of inventorying, conserving, and using biological diversity. We also need to refer to specific phylogenetic groups in studies of processes acting to generate and maintain distinctive lineages. Not all discovered monophyletic groups, at whatever level, need to or should be named: some will be trivial in evolutionary terms (i.e., of short temporal duration or marked only by minor, selectively neutral apomorphies), some cryptic (i.e., marked only by molecular or chemical apomorphies and thus nearly impossible to distinguish for practical uses), and some poorly supported (and thus subject to frequent change as more characters and taxa are discovered). There will not always be a "smallest" monophyletic group in an ontological sense; monophyletic groups exist in many organisms (especially clonal ones, but also any group with limited dispersibility) at much smaller levels than one would want to recognize formally with a Linnaean name. Thus, application of the species rank, like any other, is never automatic – it always requires independent justification.

Phylogenetic Justification

Our basic position is that there is no species problem per se in systematics. Rather, there is a taxon problem. Once one has decided what taxon names are to represent in general, then species taxa should be the same kinds of things, just the least inclusive. As discussed above, it must be recognized that there is an element of arbitrariness to the formal Linnaean nomenclatorial system. Evolution is real, as are organisms (physiological units), lineages (phylogenetic units), and demes (interbreeding units), for example. On the other hand, our classification systems are obviously human constructs, meant to serve certain purposes of our own: communication, data storage and retrieval, and predictivity. These purposes are best served by classification systems that reflect our best understanding of natural processes of evolution, and the field of systematics, in general, has settled on restricting the use of formal taxonomic names to represent phylogenetically natural, monophyletic groups. We will not repeat here the many reasons for preferring the phylogenetic approach to general-purpose classifications (see Hennig 1965, 1966; Nelson 1973; Wiley 1981; Farris 1983); instead, having accepted principles of phylogenetic classification, we will argue for the thoughtful application of these principles to the species level.

A phylogenetic systematic study of a previously unknown group of organisms involves three major temporal, logical phases. To understand the uniquely phylogenetic basis for our approach to species, it is necessary to elaborate on these phases:

1. In the precladistic phase the elements of a cladistic data matrix are assembled. These elements include OTUs (operational taxonomic units), characters, and character states. OTUs are assembled initially from grouping together of individual specimens that are homogeneous for the characters then known (see also discussion by Vrana and Wheeler 1992). Hennig (1966) himself laid this process out quite well. In his words, "the individual is to be regarded as the lowest taxonomic group category" (1966:65). In Hennig's system, the individual organism is regarded as being composed of semaphoronts (character bearers), which are basically "the individual in a certain, theoretically infinitely small, time span of its life, during which it can be considered unchangeable" (1966:65). Semaphoronts are connected by ontogenetic relationships to form the individual organism; individual organisms

are connected by tokogenetic relationships to form ancestor-descendant lineages. Hennig was quite explicit (1966:66–70) in showing that although the above ontology is clear, the empirical process of grouping individual organisms together into hypotheses of species is far from clear. This complex process involves considerable reciprocal illumination (because developing hypotheses of distinct, independent characters with discrete states goes hand in hand with developing hypotheses of homogeneous OTUs). There is no "magic bullet," no obvious, theory-free way to individuate species. The process must involve analysis, and that analysis must be explicitly phylogenetic.

2. Cladistic analysis involves translation of the data matrix into a cladogram. Reciprocal illumination is often involved here as well because incongruence between characters or odd behavior of particular OTUs may lead to a return to phase 1, a reexamination of OTUs and characters, primarily to check for fit to the assumptions of the cladistic method (i.e., that OTUs should be homogeneous for the characters used and should be the result of a diverging phylogenetic process rather than a reticulating, tokogenetic process; characters should be discrete, heritable, and independent).

3. Classifications based on an assessment of the relative support for different clades provide a basis for evolutionary studies. Formal taxa (including species) are named here on the basis of clear support for their existence as monophyletic cross-sections of a lineage and for their utility in developing and discussing process theories.

Discussion and Conclusions

Reticulation

Certain fundamental assumptions must be made in order to justify the use of cladistic parsimony for phylogenetic reconstruction. These have been discussed by a number of people (see summary by Sober 1988); Mishler (1994) argued that five basic assumptions are necessary:

1. Replication (in the sense of Hull 1980; Brandon 1990) must occur to form *lineages* (the diachronic ancestor-descendant relationship; Wilson 1995).

2. Particular features to be used as historical markers (*characters*) must have discrete variants (*character states* empirically, *transformational homologs* ontologically) that show a strong correlation (heritability in a population genetic sense) between parent and offspring.

3. Divergence (branching of lineages) must occur, as compared with reticulation, giving rise to patterns of *taxic homologs* (in the sense of Patterson 1982) shared among *sister groups* (the synchronic monophyly relationship).

4. Independence must occur among different characters; that is, no process (e.g., natural selection, gene conversion, developmental constraints) is

operating to produce nonhomologous character associations that over-
whelm taxic homologs, indicating common history.
5. Transformation in particular characters must occur at a relatively low
rate, as compared with divergence (see Mishler 1994 for discussion and
further literature references).

Note that the first and third assumptions are ontological, whereas the second,
fourth, and fifth assumptions are merely epistemological. If one of the latter
are violated to some extent, we can still get the true relationships. If the third
assumption is violated by reticulation, true relationships of the resulting hybrid
literally cannot be obtained via cladistic parsimony. Note that this is, of course,
the case with any other phylogenetic reconstruction algorithm introduced,
whether based on distances, phenetics, maximum likelihood, or some other
criterion. However, there is hope for future development of algorithms to detect
reticulation because it is possible to infer hybridity based on genomic studies
(using chromosomal markers or allelic markers such as allozymes or RAPDs
[randomly amplified polymorphic DNAs]; Rieseberg et al. 1990; Arnold et al.
1991; Rieseberg 1991; Arnold et al. 1992).

Reticulation is thus the *bête noire* for cladistics, as initially recognized by
Hennig. There are a number of different sources of homoplasy (incongruency
between certain character distributions and the cladogram based on maximum
parsimony), such as adaptive convergence, gene conversion, developmental
constraints, mistaken coding, and reticulation. The last-named factor is the most
problematical because it involves the fundamental model of reality underlying
cladistic analysis. The other factors are cases of mistaken hypotheses of homology, whereas homoplastic character distributions due to reticulate evolution
involve true homologies whose mode of transmission is not treelike.

Hennig and later Nixon and Wheeler were correct in focusing on reticulation
and the problems it causes for cladistics. Our opinion of the significance of this
problem for the species question differs to some extent from theirs, however, for
the following reasons: (1) just as barriers to reticulation are often not complete,
reticulation is not a complete barrier to cladistic analysis; and (2) reticulate relationships range from intense (in panmictic, sexually reproducing groups, where
individual relationships are exclusively reticulate) to less intense (in spatially or
temporally subdivided groups, where both reticulate and divergent relationships
exist, facultatively and/or obligatorily, among individuals).

The presence of some reticulation is not an absolute barrier to cladistic
reconstructions. We can reconstruct relationships in the face of some amount of
reticulation (how much is not yet clear, but is amenable to study). For example,
McDade (1992) has shown that incorporating a few known hybrids in an analysis
of "good" species does not seriously affect the cladistic topology of the good
species. Of course, the hybrids cannot be placed correctly in a reticulate position solely via cladistic analysis, but the relationships of the nonhybrids may
be perfectly reconstructable. McDade actually gives rules predicting what a
hybrid taxon should do in a cladistic analysis; thus, there may be a self-correct-
ing mechanism here, as there is with other sources of homoplasy; even major

convergence (e.g., in cave animals) can be uncovered via cladistic analysis. As with convergence, where the application of cladistic analysis provides the only rigorous basis we have for identifying homoplasy and thus demonstrating non-parsimonious evolution (Farris 1983), the only way we can identify reticulation on the basis of character analysis alone is through the application of cladistic parsimony, followed by the examination of homoplasy to attempt to discover its source.

Furthermore, there is no consistently clear demarcation between reticulate and branching relationships. Hybridization takes place between clades of various patristic/cladistic degrees of relatedness. There is no sharp distinction between sexually versus asexually reproducing populations in a great many organisms. Bacteria exchange genetic material in a variety of ways. Diatoms, cladocerans, and rotifers commonly undergo many asexual generations, with occasional sexual generations occurring in response to environmental change; some lineages within these groups can be obligately asexual. In many diatoms, only part of a single clonal lineage can become sexual at any given time. Other forms of reticulation occur throughout nature. Rare, high-level hybridizations may occur among very divergent lineages, such as among genera of orchids; viral-mediated lateral transfer of genetic material is suspected at much higher levels.

Thus, just as there may be no largest cladistic unit for which reticulation is impossible, there may be no smallest irreducible cladistic unit within which no further diverging phylogenetic patterns occur; ontologically speaking, we are dealing with a fractal pattern. When one looks at a lineage closely, one sees a pattern of divergence of lineages within (and some reticulation, perhaps increasingly greater, as one looks at less inclusive lineages). Asexuals are the most extreme case; cladistic structure will go down to the organism level. This fractal pattern of reticulation and branching is a severe problem for phylogenetic inference by any means. But as argued above, phenomena such as symbiosis are discovered as incongruence between organismal and character phylogenies. Massive convergence in one character system is discovered by incongruence between that system and other characters. By presuming that synapomorphy is equivalent to strict taxic homology of sister groups, cladistic analysis implies that homoplasy is explainable by all other processes, including reticulation. Lacking other information, reticulation must always be presumed to be a possible explanation for homoplasy.

Assuming we want to discover reticulation by objective means (Vrana and Wheeler 1992), it will be important to focus further attention on the problem of reticulation. Were cladistic analysis to be attempted on individuals within a panmictic group, consensus cladograms would presumably be nearly completely unresolved. This would be the correct result: there is little or no cladistic structure to reconstruct in such cases. Admittedly, however, one might still get a single most parsimonious tree even with heavily reticulating units. An unproven assumption in such cases of intense reticulation among OTUs is that there would be a disproportionate number of nearly most parsimonious trees. One might also expect to observe nonrandom distributions of homoplastic characters (concerted homoplasy) in cases of hybridization. How modes of reticulation actually affect character distributions on cladograms is a productive avenue for empirical and theoretical investigations.

This avenue reflects one of the great strengths of the direct character analysis procedure of cladistics. Methods that sum information across all characters (distance or phenetic methods) instead of treating them discretely cannot directly discover reticulation. Although direct observation of reticulation (e.g., field studies of hybridization) would indicate that cladistic analysis is inappropriate for phylogenetic inference, the presence of fixed characters at some level of grouping is neither direct nor indirect evidence for reticulation below that level. Only homoplasy may be used as indirect evidence for reticulation.

In conclusion, reticulation is not a species-specific problem. Modes of reticulation may differ and may be more or less intense in different kinds of organisms. The central difficulty remains identifying reticulation events in the midst of cladistic events. At higher levels, there seems to be wide consensus that synapomorphy can be discovered in spite of reticulation. Our Phylogenetic Species Concept, a species concept that identifies species as taxa identifiable by apomorphy, is consistent with the entire phylogenetic system and in principle is no more or less vulnerable to violation of its assumptions than is any level of phylogenetic analysis.

Asexual Reproduction

Our Phylogenetic Species Concept as defined above is clearly equally applicable to sexual and asexual organisms. This is important because many lineages exist that reproduce solely or mainly by nonsexual means. On the other hand, despite claims to the contrary, Wheeler and Platnick's Phylogenetic Species Concept is not appropriate for asexual species, in part because it lacks clearly defined ranking criteria. Cladistic relationships exist down to the individual level in asexual species. Furthermore, plesiomorphically defined groups may be clades, but they are also likely to be simply grades or even polyphyletic assemblages, as is the case for higher taxa. Thus, only apomorphic characters can identify phylogenetically natural groups in asexual species; the only applicable concept here is the Phylogenetic Species Concept (M–T).

The species situation in clonal organisms was explored in detail in a series of papers in *Systematic Botany*, introduced by Mishler and Budd (1990). First of all, despite the impression given by certain writers in the field, there is no sharp distinction between sexually and asexually reproducing organisms (as discussed above). Every degree of frequency of sex exists among populations of different species, ranging from absolute asexuality, through rare fertilization events, to panmixia. One instance of sexual recombination in a million asexual generations does not suddenly change the ontological or epistemological status of a species. Secondly, the supposed difference in phylogenetic patterns between sexually and asexually reproducing organisms does not hold up under close examination.

Mishler (1990) addressed previous predictions about the discreteness of sexual versus asexual species, using a cladistic analysis of the moss genus *Tortula* (a clade within which a spectrum of sexuality occurs, ranging from frequent sexual reproduction to total asexuality). It would be predicted under standard evolutionary theory that sexual species should be more variable than asexual species within populations (because of recombination) and less variable between populations (because of the homogenizing effect of gene flow). Therefore, species in asexual groups should be less discrete than those in sexual groups. However, measures of species distinctness, either cladistic (i.e., the number of autapomorphies) or phenetic (i.e., ordinations or analyses of variance of morphometric data), showed no particular correlation with mode of reproduction. Mishler concluded that processes other than gene flow may be responsible for species formation and maintenance even in sexual groups, a finding that has implications for speciation studies (see Budd and Mishler 1990; Mishler and Budd 1990 for further discussion).

Speciation and the Phylogenetic Species Concept

Theriot (1992) investigated patterns of speciation in relation to species concepts in a species complex of diatoms with an extremely robust fossil record. He took "phenotypically irreducible clusters" (i.e., groupings of organisms not divisible by cladistically significant characters; basically the Phylogenetic Species Concept of Cracraft 1983) as OTUs in a cladistic analysis and compared the resulting phylogeny with known ecological, stratigraphic, and biogeographic data. He concluded that three autapomorphic species each were products of evolution and probably also units participating in the evolutionary process, whereas the widespread, plesiomorphic *Stephanodiscus niagarae* is neither a product nor a unit of evolution. Thus, he cautioned against accepting the smallest phenetically recognized clusters of organisms as basic units or products of evolution.

A number of potential empirical errors can occur in analyses of species, including those conducted under our Phylogenetic Species Concept. However, there is one potential "error" (i.e., characters undiscovered at the time of analysis) for which our concept is robust with respect to other phylogenetic concepts. The diatom example again illustrates this point. Zechman et al. (1994) have begun to analyze these diatoms with molecular and morphological data, identifying cladistic structure within *S. niagarae*, further demonstrating its paraphyletic nature. An important point to be made is that even if cladistic structure could be demonstrated as real within the autapomorphic species, their interpretation as an evolutionary lineage would not be altered. However, the discovery of cladistic structure within the plesiomorphic species *S. niagarae* fundamentally shifts the view of *S. niagarae* as a natural unit to merely an aggregate of lineages. Thus, with regard to the primary goal of cladistic analysis and phylogenetic systematics, the discovery of natural groups, our Phylogenetic Species Concept applies a robust interpretation (i.e., that the identified group is monophyletic) to the discovery of new characters, whereas a concept lacking the use of apomorphy does not and cannot.

In general, our Phylogenetic Species Concept remains faithful to cladistic principles and thus is subject to exactly the same promise and problems of cladistic analysis that occur at any level. Any cladistic analysis that fails to take into account the possibility of reticulation may not be realistic. Not all lineages may have evolved apomorphic characteristics, and so they may not be identifiable through character analysis. That is, there may be monophyletic groups for which there is no direct evidence. Once again, this is a general problem for cladistic analysis and is not special to the species problem. On the other hand, if the standard assumptions of cladistic analysis are met, then our Phylogenetic Species Concept identifies natural units regardless of relationships among individuals of that unit.

REFERENCES

Arnold, M. L., C. M. Buckner and J. J. Robinson. 1991. Pollen-mediated introgression and hybrid speciation in Louisiana irises. Proc. Nat. Acad. Sci., U.S.A 88: 1398–1402.

Arnold, M. L., J. J. Robinson, C. M. Buckner and B. D. Bennet. 1992. Pollen dispersal and interspecific gene flow in Louisiana irises. Heredity 68: 399–404.

Baum, D. 1992. Phylogenetic species concepts. Trends Ecol. Evol. 7: 1–2.

Brandon, R. N. 1990. Adaptation and Environment. Princeton University Press, Princeton, NJ.

Bremer, K. and H.-E. Wanntorp. 1979. Geographic populations or biological species in phylogeny reconstruction? Syst. Zool. 28: 220–224.

Budd, A. F. and B. D. Mishler. 1990. Species and evolution in clonal organisms – summary and discussion. Syst. Bot. 15: 166–171.

Cracraft, J. 1983. Species concepts and speciation analysis. Curr. Ornith. 1: 159–187.

de Queiroz, K. and J. Gauthier. 1992. Phylogenetic taxonomy. Ann. Rev. Ecol. Syst. 23: 449–480.

Donoghue, M. J. 1985. A critique of the biological species concept and recommendations for a phylogenetic alternative. Bryol. 88: 172–181.

Eldredge, N. and J. Cracraft. 1980. Phylogenetic patterns and the evolutionary process. Columbia Univ. Press, New York.

Farris, J. S. 1983. The logical basis of phylogenetic analysis. Pages 7–36 in Advances in Cladistics, Volume 2 (Platnick, N. and V. Funk, ed.). Columbia Univ. Press, NY.

Hennig, W. 1965. Phylogenetic systematics. Ann. Rev. Entomol. 10: 97–116.

Hennig, W. 1966. Phylogenetic systematics. University of Illinois Press, Urbana.

Hull, D. L. 1980. Individuality and selection. Ann. Rev. Ecol. Syst. 11: 311–332.

Hull, D. L. 1988. Science as a process. An evolutionary account of the social and conceptual development of science. Univ. Chicago Press, Chicago.

McDade, L. A. 1992. Hybrids and phylogenetic systematics II. The impact of hybrids on cladistic analysis. Evol. 46: 1329–1346.

Mishler, B. D. 1985. The morphological, developmental, and phylogenetic basis of species concepts in bryophytes. Bryol. 88: 207–214.

Mishler, B. D. 1987. Sociology of science and the future of Hennigian phylogenetic systematics. Cladistics 3: 55–60.

Mishler, B. D. 1990. Reproductive biology and species distinctions in the moss genus *Tortula*, as represented in Mexico. Syst. Bot. 15: 86–97.

Mishler, B. D. 1994. The cladistic analysis of molecular and morphological data. American Journal of Physical Anthropology

Mishler, B. D. 1999 Getting rid of species? In R. Wilson (ed.), Species: New Interdisciplinary Essays, pp.307–315. MIT Press, Cambridge, MA.

Mishler, B. D. and R. N. Brandon. 1987. Individuality, pluralism, and the phylogenetic species concept. Biology and Philosophy 2: 397–414.

Mishler, B. D. and A. F. Budd. 1990. Species and evolution in clonal organisms – introduction. Syst. Bot. 15: 79–85.

Mishler, B. D. and M. J. Donoghue. 1982. Species concepts: a case for pluralism. Syst. Zool. 31: 491–503.

Nelson, G. 1973. Classification as an expression of phylogenetic relationships. Syst. Zool. 22: 344–359.

Nelson, G. and N. Platnick. 1981. Systematics and biogeography. Cladistics and vicariance. Columbia Univ. Press, New York.

Nixon, K. C. and Q. D. Wheeler. 1990. An amplification of the phylogenetic species concept. Cladistics 6: 211–223.

Patterson, C. 1982. Morphological characters and homology. Pages 21–74 in Problems of Phylogenetic Reconstruction (Joysey, K. A. and A. E. Friday, ed.). Academic Press, London.

Rieseberg, L. H. 1991. Homoploid reticulate evolution in *Helianthus* (Asteraceae): evidence from ribosomal genes. Amer. J. Bot. 78: 1218–1237.

Rieseberg, L. H., R. Carter and S. Zona. 1990. Molecular tests of the hypothesized hybrid origin of two diploid *Helianthus* species (Asteraceae). Evolution 44: 1498–1511.

Rosen, D. E. 1978. Vicariant patterns and historical explanation in biogeography. Syst. Zool. 27: 159–188.

Rosen, D. E. 1979. Fishes from the upland and intermontane basins of Guatemala: revisionary studies and comparative geography. Bull. Amer. Mus. Nat. Hist. 162: 267–376.

Sober, E. 1988. Reconstructing the Past. MIT Press, Cambridge, MA.

Theriot, E. 1992. Clusters, species concepts, and morphological evolution of diatoms. Syst. Biol. 41: 141–157.

Vrana, P. and W. Wheeler. 1992. Individual organisms as terminal entities: laying the species problem to rest. Cladistics 8: 67–72.

Wiley, E. O. 1981. Phylogenetics: the theory and practice of phylogenetic systematics. John Wiley and Sons, New York.

Wilson, B. E. 1995. A (not-so-radical) solution to the species problem. Biol. Phil. 10: 339–356,

Zechman, F. W., E. A. Zimmer and E. C. Theriot. 1994. Use of ribosomal DNA internal transcribed spacers for phylogenetic studies in diatoms. J. Phycol. 30: 507–512.

Part II

What Should Happen to Taxa
at the Traditional Species
Level under a Rankless
Code of Nomenclature?

4 General Principles of Rankless Classification Extended to the Species Rank

Calls for the reformation of the Linnaean hierarchy have been building since the early 1990's (e.g., De Queiroz & Gauthier 1992, 1994). These authors emphasized that the existing classification system is based on a non-evolutionary world-view; the roots of the Linnaean hierarchy are in a specially-created world-view. Perhaps the idea of fixed, comparable ranks made some sense under the view that classification was reflecting God's plan with its nested elements, but under an evolutionary world view, the ranks just don't make sense. They are a anachronism causing confusion in the modern world. There are several problems with the current nomenclatorial systems (*International Code of Nomenclature for Algae, Fungi, and Plants*, Turland et al. 2018; and the *International Code of Zoological Nomenclature*). Here are some of them:

1. The current system, using a single type for a name, cannot precisely name a clade, e.g., you may name a family based on a certain type specimen, and even if were clear about what node you meant to name in your original publication, the exact phylogenetic application of your name would not be clear subsequently, after new clades are added to the picture.
2. There are not nearly enough ranks to name the thousands of levels of monophyletic groups in the tree of life. Therefore people have been increasingly using informal rankless names for higher-level nodes, but without any clear, formal specification of which clade is meant.
3. The ranks can lead to instability of usage. When a change in relationships is discovered, names may need to be changed of groups whose circumscription has not changed at all, e.g., when it was detected that the Cactaceae is nested inside of the Portulacaceae, one of these well-known family names had to be abandoned. This instability is particularly an issue at the species level, because of the binomial which links names at two different ranks. Splitting a genus into segregate genera results in many frivolous changes to species names without any change in postulated circumscription of them.
4. Groups given the same rank across biology are not comparable in any way (i.e., in age, size, amount of divergence, diversity within, etc.), but many users do not know this and employ numbers of taxa at a particular rank as an erroneous measure of diversity (Miller et al. 2018). In this way, formal ranks can lead to bad science.

For all these reasons and more the *PhyloCode* was developed and has just been published (Cantino & de Queiroz 2020). Linnaean ranks are abandoned and instead, names of clades are regarded as hierarchically nested uninomials. A clade retains its name regardless of where new knowledge might change its phylogenetic position, thus increasing nomenclatorial stability. Two or more types (called "specifiers") are used, for an efficient and accurate representation of phylogenetic relationships. Furthermore, since clade names are presented to the community without attached ranks, users are encouraged to look at the actual attributes of the clades they compare, thus improving research in comparative biology.

Despite all these advances, the species rank remains a controversial topic, even among *PhyloCode* advocates. Some (primarily zoologists) want to retain it as essentially one fundamental rank in an otherwise rankless system, others (primarily botanists) want to get rid of it entirely. The current *PhyloCode* retains the use of species-rank taxa in several important ways and the debate about removing the mentions of species from the *PhyloCode* will no doubt continue to rage (e.g., Cellinese, Baum, & Mishler 2012, reprinted below).

My view continues to be that the principles of rankless classification should be applied to terminal taxa, including the former species rank (Mishler 1999, 2010, reprinted below). Names of clades (including the terminal level), should be hierarchically nested uninomials regarded as proper names (as at all levels in the *PhyloCode*). See Fig. 4.1, which illustrates how the *PhyloCode* could be applied in a taxonomic revision at the level formerly known as species (adapted from Fisher 2006).

Despite the taxonomic instability that is caused by using binomial names for species in the current codes (discussed above), many feel there is value in a binomial name for species, in that having a higher level name (genus) to help place the lower-level name (species). However, this perceived need is covered in the *PhyloCode*. A named clade at any level intrinsically has many higher-level names available to place it, available in the database RegNum (https://www.phyloregnum.org), sometimes called the "clade address." If you are writing a general paper for readers who are not familiar with fine-scale classification in a group, you would include a higher level clade name to orient them (which is current practice now, actually). If I am talking about my favorite study moss (currently called *Syntrichia caninervis* as a binomial) in a non-specialist context I might write "*Caninervis* (*Syntrichia*)" or even "*Caninervis* (*Syntrichia, Musci*)" if writing in a very general context where the reader might not be expected to know *Syntrichia* is a moss.

At any one time, there will be a least-inclusive named clade, which can be called the **S**mallest **Na**med and **R**egistered **C**lades or SNaRCs (Mishler & Wilkins 2018, reprinted below). SNaRC is not a rank. There is no implication that the SNaRCs are necessarily comparable in any way other than that they are hypothesized to be clades like all other taxa named under the *PhyloCode*. There is also no implication that SNaRCs are the smallest clades in truth – they are just the smallest clades that a taxonomist wants to give a formal, registered name to in practice, i.e., the least inclusive clades that taxonomist feels comfortable naming, given current data.

The tree of life is inherently fractal. Look closely at one branch (lineage) of a phylogeny and it dissolves into many separate lineages, and so on down to a very fine scale. Thus clades occur at a very large number of nested levels. All are interesting

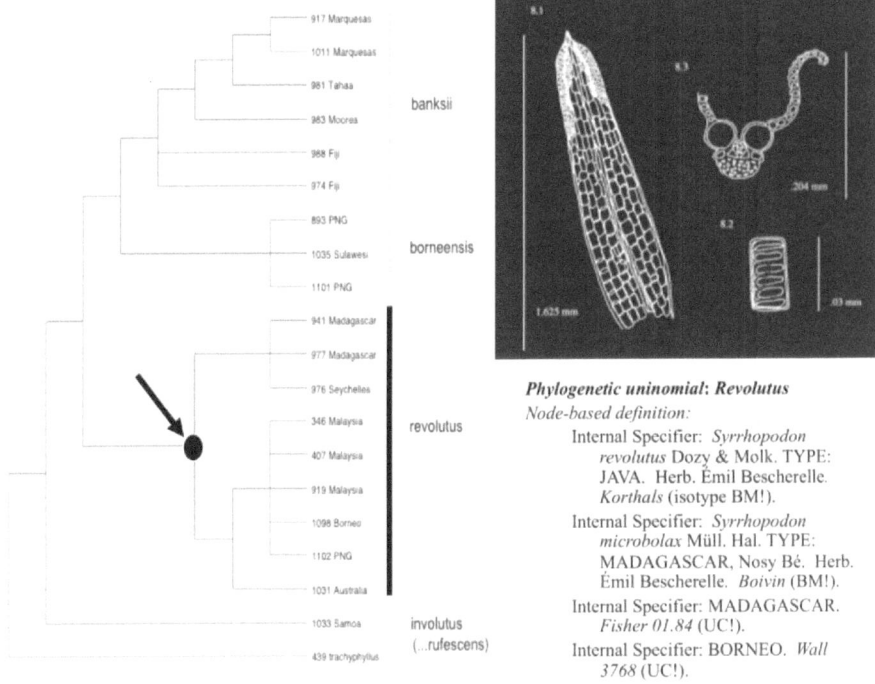

Phylogenetic uninomial: Revolutus
Node-based definition:
    Internal Specifier: *Syrrhopodon*
       revolutus Dozy & Molk. TYPE:
       JAVA. Herb. Émil Bescherelle.
       Korthals (isotype BM!).
    Internal Specifier: *Syrrhopodon*
       microbolax Müll. Hal. TYPE:
       MADAGASCAR, Nosy Bé. Herb.
       Émil Bescherelle. *Boivin* (BM!).
    Internal Specifier: MADAGASCAR.
       Fisher 01.84 (UC!).
    Internal Specifier: BORNEO. *Wall*
       3768 (UC!).

FIGURE 4.1   Example of the naming of a clade at the level formerly known as species. The specifiers – similar to types in the current ranked systems of nomenclature, but with multiple specimens used to triangulate to a definitive node – are a mix of current type specimens and new collections. The moss species named *Syrrhopodon revolutus* under the current code is a SNaRC named *Revolutus* under this *PhyloCode*-like description. However, note that the *PhyloCode* currently does not allow this type of conversion at the species rank, a unique and inconsistent exception to its basic principles which is a matter of considerable controversy (see text for further discussion, especially Cellinese et al. 2012 reprinted later in this chapter). (Modified from Fisher 2006, *Systematic Botany, 31*, 13–30.)

for certain questions, none are "fundamental." Therefore to study biodiversity, its evolutionary origins and ecological interactions, the way forward is to study the whole tree and carefully choose clades to compare at the level appropriate to the question being asked. I'll expand on this theme in Chapter 5.

LITERATURE CITED

Cantino, P.D., and K. de Queiroz. 2020. *International Code of Phylogenetic Nomenclature (PhyloCode)*. CRC Press, Boca Raton, Florida.
Cellinese, N., D.A. Baum, and B.D. Mishler. 2012. Species and phylogenetic nomenclature. *Systematic Biology* 61: 885–891.
de Queiroz, K. and J. Gauthier. 1992. Phylogenetic taxonomy. *Annual Review of Ecology and Systematics* 23: 449–480.
de Queiroz, K. and J. Gauthier 1994. Toward a phylogenetic system of biological nomenclature. *Trends in Ecology & Evolution* 9: 27–31.

Fisher, K.M. 2006. Rank-free monography: a practical example from the moss clade *Leucophanella* (Calymperaceae). *Systematic Botany* 31: 13–30.

Miller, J.T., G. Jolley-Rogers, B.D. Mishler, and A.H. Thornhill. 2018. Phylogenetic diversity is a better measure of biodiversity than taxon counting. *Journal of Systematics and Evolution* 56: 663–667.

Mishler, B.D. 1999. Getting rid of species? Pp. 307–315 in: *Species, New interdisciplinary essays*, R. A. Wilson (ed.). Bradford/MIT Press, Cambridge, MA.

Mishler, B.D. 2010. Species are not uniquely real biological entities. Pp. 110–122 in: *Contemporary Debates in Philosophy of Biology*, F. Ayala and R. Arp (eds.). Wiley-Blackwell, Weinheim, Germany.

Mishler, B.D. and J.S. Wilkins. 2018. The hunting of the SNaRC: a snarky solution to the species problem. *Philosophy, Theory, and Practice in Biology* 10: 1–18.

Turland, N.J., J.H. Wiersema, F.R. Barrie, W. Greuter, D.L. Hawksworth, P.S. Herendeen, S. Knapp, W.-H. Kusber, D.-Z. Li, K. Marhold, T.W. May, J. McNeill, A.M. Monro, J. Prado, M.J Price and G.F. Smith (eds.), 2018. *International Code of Nomenclature for algae, fungi, and plants (Shenzhen Code) adopted by the Nineteenth International Botanical Congress Shenzhen, China, July 2017.* Regnum Vegetabile 159. Koeltz Botanical Books, Oberreifenberg, Germany.

Getting Rid of Species?[14]

Abstract

This paper explores the implications of generalizing the species problem as a special case of the taxon problem. Once a decision is made about what taxa, in general, are to represent, then species, in particular, are simply the least inclusive taxon of that type. As I favor a Hennigian phylogenetic basis for taxonomy, I explore the application of phylogenetics to species taxa. Furthermore, I advocate a novel extension of the recent calls for rank-free phylogenetic taxonomy to the species level. In brief, the argument is: (1) Species must be treated as just one taxon among many; (2) All taxa should be monophyletic groups; (3) Because of problems with instability and lack of comparability of ranks in the formal Linnaean system, we need to move to a rank-free formal classification system; (4) In such a system, not all hypothesized monophyletic groups need be named, but those that are named formally should be given unranked (but hierarchically nested) uninomials; (5) The least inclusive taxon, formally known as "species," should be treated in the same unranked manner. Finally, I explore the practical implications of eliminating the rank of species for such areas as ecology, evolution, and conservation, and conclude that these purposes are better served by this move.

The debate about species concepts over the last 20 years follows a curious pattern. Rather than moving towards some kind of consensus, as one might expect, the trend has been towards an ever-increasing proliferation of concepts. Starting with the widely accepted species concept that had taken over in the 1940's and 1950's as a result of the Modern Synthesis, the Biological Species Concept, we heard calls for change from botanists, behaviorists, etc. Despite the babel of new concepts, the BSC continues to have fervent advocates (Avise and Ball, 1990; Avise and Wollenberg, 1997) and has itself spawned several new variants. A recent paper by Mayden (1997) lists no fewer than 22 prevailing concepts! It appears we can never eliminate any existing concept, only produce new ones.

Why is this? The obvious conclusion one might draw, that biologists are contrarians who each want to make their own personal mark in a debate and thus coin their own personal concept to defend, is really not the case – this is no debate about semantics. The conceptual divisions are major, and real. In my opinion, the plethora of ways in which different workers want to use the species category reflects an underlying plethora of valid ways of looking at biological diversity. The way forward is to recognize this, and face the implications: the basis of this confusion over species concepts is a result of heroic but doomed attempts to shoehorn all this variation into an outdated and misguided

[14]B.D. Mishler. 1999. Getting rid of species? In R. Wilson (ed.), Species: New Interdisciplinary Essays, pp.307–315. MIT Press. [reprinted by permission]

classification system, the ranked Linnaean hierarchy. Most of the confusion can be removed simply by removing the ranks. The issues that remain can then be dealt with by carefully considering what we want formal classification to represent as the general reference system, and then carefully specifying criteria for grouping organisms into these formal classifications.

To develop this argument, I will first make the case for generalizing the species problem as a special case of the taxon problem. For a consistent general reference classification system, all taxa must be of the same type; species should be regarded as simply the least-inclusive taxon in the system. Then I will review the reasons for why phylogeny provides the best basis for the general-purpose classification; species should thus be considered as just another phylogenetically-based taxon. Next, I will address the recent calls for rank-free classification in general, and pursue the central thesis of this paper: *the species rank must disappear along with all the other ranks*. Finally, I will explore the practical implications of eliminating the rank of species for such areas as ecology, evolution, and conservation.

Species as Just Another Taxon

Many authors have made a firm distinction, in their particular theories of systematics, between species and higher taxa (e.g., Wiley, 1981; Nelson and Platnick, 1981; Nixon and Wheeler, 1990); see discussion by De Quieroz, this volume). The idea is that somehow species are units directly participating in the evolutionary process, while higher taxa are at most lineages resulting from past evolutionary events. However nicely drawn this distinction is in theory, these arguments have resulted more from wishful thinking than from empirical observations. When anyone has looked closely for an empirical criterion to uniquely and universally distinguish the species rank from all others, the attempt has failed.

One early suggestion was phenetic: a species is a cluster of organisms in Euclidean space separated from other such clusters by some distinct and comparable gap (e.g., Levin 1979). This has been clearly shown to be mistaken – phenotypic clusters are actually nested inside each other with continuously varying gap sizes. Current entities ranked as species are not comparable either in the amount of phenotypic space they occupy or the size of the "moat" around them, nor can they be made to be comparable through any massive realignment of current usage.

Another suggestion for a unique ranking criterion for species is expressed in the biological species concept: a species is a reproductive community separated by a major barrier to crossing with other such communities (Mayr, 1982). Like the phenetic gap, this view (nice in theory perhaps) fails when looking at real organisms. Despite the publication of many conceptual diagrams that depict a distinct break between reticulating and divergent relationships at some level (Nixon and Wheeler, 1990; Roth, 1991; Graybeal, 1995) actual data suggests that in most groups the probability of intercrossability decreases gradually as you compare more and more inclusive groups (Mishler and Donoghue 1982; Maddison 1997). There usually is not a distinct point at which the possibility of reticulation drops precipitously to zero.

Similar suggestions have been made based on ecological criteria: a species is a group of organisms occupying some specific and unitary ecological niche (Van

Valen, 1976). May be species "can define themselves" – we just need to see whether two organisms treat each other as belonging to the same or different species. Again, actual studies show no such distinctive level where ecological interactions change abruptly from "within-kind" to "between kind." There can be cryptic, ecologically distinct groups below the species level, and large guilds of organisms from divergent species acting as one group ecologically in some situations.

Finally, there have been attempts to distinguish species from all other taxa phylogenetically (Nixon and Wheeler 1990, Graybeal 1995, Baum, 1992). In this view, species are the smallest divergent lineage – inside of which there is no recoverable divergent phylogenetic structure (only reticulation). Again, nice in theory, but unsound empirically, at least as a general principle (Mishler and Donoghue, 1982; Mishler and Theriot, in press). Some biological situations fit the model well (e.g., in organisms with complex and well-defined sexual mate recognition systems and no mode of asexual propagation). However, in many clonal groups (e.g., aspen trees, bracken fern) there are discernible lineages going down to the within-organism level (the problem of "too little sex"; Templeton, 1989). On the other hand, there are occasional horizontal transfer events ("reticulation") between very divergent lineages (the problem of "too much sex"; Templeton, 1989). In all such cases, there is a large gray area between strictly diverging patterns of gene genealogies and strictly recombining ones (contra Avise and Wollenberg, 1997).

To sum up, there is no (and is unlikely to be any) criterion for distinguishing species from other ranks in the Linnaean hierarchy. This is not to say that particular species taxa are unreal; they are, but only in the sense that taxa at all levels are. Species are not special.

The Necessity for Phylogenetic Classifications

The debate over classification has a long and checkered history. This is not the place to detail this history fully (see Stevens, 1994; and the chapter by Ereshevsky, this volume). I want to begin with the conceptual upheaval in the 1970s and 1980s that resulted in the ascension of Hennigian Phylogenetic Systematics (for a detailed treatment see the masterful book by Hull, 1988). Many issues were at stake in that era, foremost of which was the nature of taxa. Are they just convenient groupings of organisms with similar features, or are they lineages, marked by homologies? A general, if not completely universal consensus has been reached, that taxa are (or at least should be) the latter (Hennig, 1966; Nelson, 1973; Farris, 1983; Sober, 1988).

A full review of the arguments for why formal taxonomic names should be used solely to represent phylogenetic groups is beyond the scope of this paper, but they can be summarized as follows: evolution is the single most powerful and general process underlying biological diversity. The major outcome of the evolutionary process is the production of an ever-branching phylogenetic tree, through descent with modification along the branches. This results in life being organized as a hierarchy of nested monophyletic groups. Since the most effective and natural classification systems are those that "capture" entities resulting from processes generating the things being classified, the general biological classification system should be used to reflect the tree of life.

The German entomologist Willi Hennig codified the meaning of these evolutionary outcomes for systematics, in what has been called the Hennig Principle
(Hennig, 1965, 1966). Hennig's seminal contribution was to note that in a system
evolving via descent with modification and splitting of lineages, characters that
changed state along a particular lineage can serve to indicate the prior existence
of that lineage, even after further splitting occurs. The "Hennig Principle" follows from this: homologous similarities[15] among organisms come in two basic
kinds, synapomorphies due to immediate shared ancestry (i.e., a common
ancestor at a specific phylogenetic level), and symplesiomorphies due to more
distant ancestry. Only the former are useful for reconstructing the relative order
of branching events in phylogeny. A corollary of the Hennig Principle is that
classification should reflect reconstructed branching order; only monophyletic
groups[16] should be formally named. Phylogenetic taxa will thus be "natural" in
the sense of being the result of the evolutionary process.

This isn't to say that phylogeny is the only important organizing principle
in biology, There are many ways of classifying organisms into a hierarchy,
because of the many biological processes impinging on organisms. Many kinds
of non-phylogenetic biological groupings are unquestionably useful for special
purposes (e.g., "producers," "rain forests," "hummingbird pollinated plants,"
"bacteria"). However, it is generally agreed that there should be one consistent,
general-purpose, reference system, for which the Linnaean hierarchy should be
reserved. Phylogeny is the best criterion for the general-purpose classification,
both theoretically (the tree of life is the single universal outcome of the evolutionary process) and practically (phylogenetic relationship is the best criterion for summarizing known data about attributes of organisms and predicting
unknown attributes). The other possible ways to classify can of course be used
simultaneously, but should be regarded as special-purpose classifications and
clearly distinguished from phylogenetic formal taxa.

The Advantages of a Rank-Free Taxonomy

A number of recent calls have been made for the reformation of the Linnaean
hierarchy (e.g., De Queiroz & Gauthier, 1992). These authors have emphasized
that the existing system is based on a non-evolutionary world-view; the roots of
the Linnaean hierarchy are a specially-created world-view. Perhaps the idea of
fixed ranks made some sense under that view, but under an evolutionary world
view, they don't make sense. Most aspects of the current code, including priority,
revolve around the ranks, which leads to instability of usage. For example, when a
change in relationships is discovered, several names often need to be changed to
adjust, including those of groups whose circumscription has not changed. Frivolous
changes in names often occur when authors merely change the rank of a group
without any change in postulated relationships. While practicing systematists

---

[15] In Hennigian phylogenetic systematics, "homology" is defined historcally as a feature shared by
two organisms because of descent from a common ancestor that had that feature.

[16] A strictly monophyletic group is one that contains all and only descendants of a common ancestor. A paraphyletic group is one the excludes some of the descendants of the common ancestor.

know that groups given the same rank across biology are not comparable in any way (i.e., in age, size, amount of divergence, diversity within, etc.), many users do not know this. For example, ecologists and macroevolutionists often count numbers of taxa at a particular rank as an erroneous measure of "biodiversity." The nonequivalence of ranks means that at best (to those who are knowledgeable) they are a meaningless formality and perhaps not more than a hindrance. At worst, in the hands of a user of classifications who naively assumes groups at the same rank are comparable in some way, formal ranks lead to bad science.

It is not completely clear at this point how exactly a new code of nomenclature should be written, but the basics are clear. Such a new code should maintain the principle of priority (the first name for a lineage should be followed) and other aspects of the current code that promote effective communication of new names to the community. However, the major change would be that the Linnaean ranks should be abandoned, for efficient and accurate representation of phylogenetic relationships. Instead, names of clades should be hierarchically nested uninomials regarded as proper names. A clade would retain its name regardless of where new knowledge might change its phylogenetic position, thus increasing nomenclatorial stability. Furthermore, since clade names would be presented to the community without attached ranks, users would be encouraged to look at the actual attributes of the clades they compare, thus improving research in comparative biology.

It is important to emphasize that, despite misrepresentations to the contrary that have appeared, those who advocate getting rid of ranks don't at all advocate getting rid of the hierarchy in biological classification. Nesting of groups within groups is essential because of the tree-like nature of phylogenetic organization. Think of a non-systematic example – a grocer might classify table salt as a spice and group spices together under the category "food items." This simple hierarchy is clear but requires no ranks to be understood. In fact, all human thought is organized into hierarchies, and becoming educated in a field is essentially learning the hierarchical arrangement of concepts in that field. Taxonomy is unusual in the assigning of named ranks to its hierarchies – there are superfluous to true understanding.

Getting Rid of the Species Rank

Curiously, so far in this debate, even the advocates of rank-free phylogenetic classification have retained the species rank as a special case. All other ranks are to be abandoned, but the species rank kept, probably because the species concept is so ingrained and comfortable in current thinking. However, all the arguments that can be massed against Linnaean ranked classification can be brought to bear against the species rank as well. As difficult as it is to overthrow ingrained habits of thinking, logical consistency demands that all levels in the classification should be treated alike.

Given the background developed in the three previous sections, the conclusion seems inescapable. The species rank must go the way of all others. We must end the endless bickering over how this rank should be applied, and instead, get rid of the rank itself. This is truly the "radical solution to the species problem" sought unsuccessfully by Ghiselin. Biological classification should be a set of nested, named groups for internested clades. Not all clades need to be named, but those that are should be named on the basis of evidence for monophyly (see further discussion of the meaning of monophyly in Mishler and Brandon, 1987). We stop naming groups at some point approaching the tips of the phylogeny because we don't have solid evidence for monophyly at the present stage of knowledge. This may be due to rampant reticulation going on below some point, or simply lack of good markers for distinguishing finer clades. But, we shouldn't pretend that the smallest clades named at a particular time are ontologically different from other, more inclusive named clades. Further research could easily result in the subdividing of these groups, or the lumping of several of them into one if their original support is discovered to be faulty.

Given the redundancy now present in species epithets (e.g., "californica" is used in many genera), there needs to be a way to uniquely place each smallest named clade in the classification. My recommendation for nomenclature at the least inclusive level under a totally rank-free classification would be to regard names in a similar way as personal names are regarded in Arabic culture. Each clade, including the least inclusive one named, has its own uninomial name; however, the genealogical relationships of a clade are preserved in a polynomial giving the lineage of that clade in higher and higher groups. So, the familiar binomial, which does after all present some grouping information to the user, could be retained but should be inverted. Our own short clade name thus should be <u>Sapiens Homo</u>. The full name for our terminal clade should be regarded as a polynominal giving them more and more inclusive clades names all the way back. To use the human example, this would be something like <u>Sapiens Homo Homidae Primate Mammalia Vertebrata Metazoa Eucaryota Life</u>.[17] But again as in a traditional Arabic name this formal and complete name would only be used rarely and for the most formal purposes (but would be very useful behind the scenes for data-basing purposes) – the everyday name of the clade would be <u>Sapiens Homo</u>.

Practical Implications

"Getting rid of species" has another, all too ominous meaning in today's world. Named species are being driven to (and over) the brink of extinction at a rapid rate. What will be the implications of the view of taxa advocated in this paper? If we get rid of the species rank, with all its problems, will we hamstring conservation efforts? I tend to think not; scientific honesty seems the best policy here as elsewhere, The rather mindless approach that has been followed in conservation, that if a lineage is ranked as a species it is worth saving, while if it is not considered a species it is not worth considering, is misguided in many ways. It is wrong scientifically as discussed above; the species rank is a human judgment rather than any objective point along the trajectory of diverging lineages. It is also wrong ethically; any recognizable lineage is worth conservation consideration.

---

[17] Note that some of the nested clades will have a formal suffix indicating their previous rank (e.g., "-idae" for family). While these ending would be retained for exiting clade names, in order to avoid confusion, there would be no meaning attached to them and newly proposed clade names would have no particular suffix requirement.

Not all lineages need be conserved, or at least given the same conservation priority, but such judgments should be made on a case by case basis.

All biologists are concerned about defining biodiversity and about its current plight, thus the radical move suggested here (getting rid of the species rank) will no doubt concern many. A common response to this move, from those who want to characterize and conserve biodiversity, involves a complaint that "without species, we will have no way of quantifying biodiversity or of convincing people to preserve it." This viewpoint, while expressing commendable and important concerns, is ultimately misguided, both in theoretical and practical terms. There may a comfortable self-deception going on to the contrary, but only a moment of thoughtful reflection is enough to realize that species are not comparable in any important sense, and cannot be made so.

However, the recognition that a count of species is not a good measure of biodiversity, does not mean that biodiversity cannot be quantified. All named species are unique, with their own properties and features, and represent only the tips of the underlying iceberg of biodiversity. We must face these facts and move to develop valid measures of the diversity of lineages taking into account their actual properties and phylogenetic significance. A number of workers have suggested quantitative measures for phylogenetic biodiversity, which take into account the number of branch points, and possibly branch lengths, separating the tips on the tree (Vane-Wright et al., 1991; Faith, 1992a,b).

Many macroevolutionary studies are framed in terms of comparing diversity patterns at some particular rank (e.g., families of marine invertebrates, phyla of animals). The adoption of rank-free classification would (fortunately) make such studies impossible, but would it make all studies of macroevolution impossible? Of course not – comparisons among clades would still be quite feasible, but it would be up to the investigator to establish that the clades being compared were the same with respect to the necessary properties (i.e., equivalent age or disparity, etc.). Similar arguments could be made with respect to the many ecological studies that compare numbers of species in different regions or communities. The bottom line is that rank-free classification would lead to much more accurate research in ecology and evolution. This is because, instead of relying on equivalence in taxonomic rank as a (very) crude proxy for comparability of lineages, investigators would be encouraged to use cladograms directly in their comparative studies. Given the rapid progress in development of quantitative comparative methods (Funk and Brooks, 1990; Brooks and McLennan, 1991; Harvey and Pagel, 1991; Martins, 1996), and the rapid proliferation of ever-improving cladograms for most groups of organisms, this can only be for the best.

REFERENCES

Avise, J. C. and R. M. Ball. 1990. Principles of genealogical concordance in species concepts and biological taxonomy. Oxford Surv. Evol. Biol. 7: 45–67.

Avise, J. C. and K. Wollenberg. 1997. Phylogenetics and the origin of species. Proc. Nat. Acad. Sci., U.S.A 94: 7748–7755.

Baum, D. 1992. Phylogenetic species concepts. Trends Ecol. Evol. 7: 1–2.

Brooks, D. R. and D. A. Mclennan. 1991. Phylogeny, ecology, and behavior. Univ. Chicago Press, Chicago.

de Queiroz, K. and J. Gauthier. 1992. Phylogenetic taxonomy. Ann. Rev. Ecol. Syst. 23: 449–480.

Faith, D. P. 1992a. Conservation evaluation and phylogenetic diversity. Biol. Conserv. 61:1–10.

Faith, D. P. 1992b. Systematics and conservation: on predicting the feature diversity of subsets of taxa. Cladistics 8:361–373.

Farris, J. S. 1983. The logical basis of phylogenetic analysis. Pages 7–36 in Advances in cladistics, Volume 2 (Platnick, N. and V. Funk, eds.). Columbia Univ. Press, NY.

Funk, V. A. and D. R. Brooks. 1990. Phylogenetic systematics as the basis of comparative biology. Smithsonian Institution Press, Washington, D.C.

Graybeal, A. 1995. Naming species. Syst. Biol. 44: 237–250.

Harvey, P. H. and M. D. Pagel. 1991. The comparative method in evolutionary biology., New York University Press, New York.

Hennig, W. 1965. Phylogenetic systematics. Annu. Rev. Entomol. 10: 97–116.

Hennig, W. 1966. Phylogenetic systematics. University of Illinois Press, Urbana.

Hull, D. L. 1988. Science as a process: an evolutionary account of the social and conceptual development of science. Univ. Chicago Press, Chicago.

Levin, D. A. 1979. The nature of plant species. Science 204: 381–384.

Maddison, W. P. 1997. Gene trees in species trees. Syst. Biol. 46: 523–536.

Martins, E. P. 1996. Phylogenies, spatial autoregression, and the comparative method: a computer simulation test. Evolution 50: 1750–1765.

Mayden, R. L. 1997. A hierarchy of species concepts: the denouement in the saga of the species problem. Pages 381–424 in Species: the units of biodiversity (Claridge, M. F., H. A. Hawah and M. R. Wilson, eds.). Chapman and Hall, London.

Mayr, E. 1982. The growth of biological thought. Harvard University Press, Cambridge, Mass.

Mishler, B. D. and R. N. Brandon. 1987. Individuality, pluralism, and the phylogenetic species concept. Biol. Philos. 2: 397–414.

Mishler, B. D. and M. J. Donoghue. 1982. Species concepts: a case for pluralism. Syst. Zool. 31: 491–503.

Mishler, B.D. and E. Theriot. In Press. Monophyly, apomorphy, and phylogenetic species concepts. Three chapters in Q. D. Wheeler & R. Meier (eds.), Species concepts and phylogenetic theory: adebate. Columbia University Press, New York.

Nelson, G. 1973. Classification as an expression of phylogenetic relationships. Syst. Zool. 22: 344–359.

Nelson, G. and N. Platnick. 1981. Systematics and biogeography. Cladistics and vicariance. Columbia Univ. Press, New York.

Nixon, K. C. and Q. D. Wheeler. 1990. An amplification of the phylogenetic species concept. Cladistics 6: 211–223.

Roth, V. L. 1991. Homology and hierarchies: problems solved and unresolved. J. Evol. Biol. 4: 167–194.

Stevens, P. F. 1994. The development of biological systematics. Columbia University Press, New York.

Sober, E. 1988. Reconstructing the past. MIT Press, Cambridge, MA.

Templeton, A. R. 1989. The meaning of species and speciation: a genetic perspective. in Speciation and its consequences (Otte, D. and J. A. Endler, ed.). Sinauer Associates, Sunderland, Mass.

Vane-Wright, R. I., C. J. Humphries And P. H. Williams. 1991. What to protect? –systematics and the agony of choice. Biol. Conserv. 55:235–254.

Van Valen, L. 1976. Ecological species, multi-species, and oaks. Taxon 25: 233–239.

Wiley, E. O. 1981. Phylogenetics: the theory and practice of phylogenetic systematics. John Wiley and Sons, New York.

Species are not Uniquely Real Biological Entities[18]

Abstract

Are species uniquely real biological entities? This question is one of the most controversial topics today in such areas of biology as ecology, systematics, conservation, population genetics, and evolution. "Species" currently play a central role in both theory and practice in these areas, and have a large place in the public's perception of biological diversity as well. This question can be decomposed into two parts: (1) Are species real, and in what sense? (2) If real, is their reality the same as entities smaller or larger than them – i.e., are they real in a sense that genera or subspecies are not? This paper will briefly review historical and current opinions on these questions, but will primarily advocate one particular position that appears to fit biological reality as now understood: that species properly defined *are* real entities, but not *uniquely* real. The longstanding "species problem" can be solved by realizing that there is no such thing as species after all! The so-called "species problem" is really just a special case of the taxon problem. Once a decision is made about what taxa, in general, are to represent, then those groups currently known as species are simply the least inclusive taxa of that type. As I favor a phylogenetic basis for taxonomy, I want to look at how to include terminal taxa in the PhyloCode, currently a controversial topic even among PhyloCode supporters. In brief, my argument is: (1) life is organized in a hierarchy of nested monophyletic groups – some of them quite fine-scale, well below the level we currently call species; (2) not all known monophyletic groups need be named, just the ones that are important to process or conservation studies and that have good support; (3) those that are named taxonomically should be given unranked (but hierarchically nested) uninomials; and (4) formal ranks, including species, should be abandoned. I will conclude with a brief discussion of the implication of my position on species for academic studies in ecology and evolution as well as for practical applications in biodiversity inventories and conservation biology.

"But be warned, you who thirst for knowledge, be warned about the thicket of opinions and the fight over words." Hermann Hesse, *Siddhartha*

Historical and Current Views of Species

Over the history of science, people have taken a number of different positions on these issues involving the reality of species. The fundamental view throughout

---

[18] B.D. Mishler. 2010. Species are not uniquely real biological entities. In F. Ayala and R. Arp (eds.), Contemporary Debates in Philosophy of Biology, pp. 110–122. Wiley-Blackwell. [reprinted by permission]

the classical period (basically from the ancient Greeks until Darwin) was that species are indeed the basic, real units of life. The basis for their reality was initially conceived of in a typological or idealistic framework; species were viewed as the basic kinds or types of living things. Later, under the influence of Christian theology, the basis for the fundamental reality of species changed somewhat. They were still viewed as basic kinds, but now as specially created "ideas" in the mind of the creator. Taxonomic groups at more inclusive levels were also manifestations of the creator's ideas, but species were the fundamental kinds, the building blocks of life. This idea continues to the present in the attitudes of the majority of the general public in the United States, under the influence of creationism.

The course of science took a somewhat different path than the public view. The Darwinian revolution did not question the reality of species in scientists' thinking (although see below for Darwin's contribution to a shift in thinking about their uniqueness), but changed the perception of their nature greatly. Instead of representing a natural kind defined by certain necessary and sufficient characteristics, species came to be seen as a natural genealogical unit composed of organisms historically related to each other, with a beginning and an end, *not* defined by any characteristics (i.e., "individuals" in the philosophical sense; Ghiselin, 1974; Hull, 1978; Mishler & Brandon, 1987). They were viewed as a cross-section of a lineage (de Queiroz, 1999). In the Modern Synthesis (called such at the time, but looking rather dated these days!), a view solidified of species being the largest group of interbreeding organisms (the gene pool) and as such the most fundamental unit in which evolutionary change takes place (*the biological species concept*; BSC; Mayr, 1942, 1982). Species came to be regarded as a fundamental level in the hierarchy of biological organization (e.g., molecule, cell, tissue, organism, population, species, community, ecosystem).

This view was nearly unanimous until the 1960s, when, under the influence of highly empirical operationalist philosophies of science then in style, and the seemingly "objective" application of computer algorithms to science, an approach emerged called "numerical taxonomy" or "phenetics." In this view, taxa at all levels, including species, were viewed in a nominalistic manner. A species was just a cluster of similar organisms grouped at some arbitrary numerical level of similarity (*the phenetic species concept*; Levin, 1979; Sokal & Crovello, 1970). It was considered to be unnecessary and wrong-headed to require anything about a deeper basis for reality, whether relatedness or interbreeding ability, to describe species. If named species later turned out to be something useful for inferences about evolutionary or ecological processes, then fine, but their recognition as species was best kept separate from process considerations.

One trend apparent in the history of thinking about species has to do with organismal specialty; to a large extent, there has been a sociological difference among communities of systematists studying different kinds of organisms. Zoologists tended to favor the biological species concept (Coyne, Orr, & Futuyma, 1988), while botanists and bacteriologists tended to favor the phenetic species concept (e.g., Levin, 1979; Sokal & Crovello, 1970). There have been some exceptions: for example, Grant (1981), Rieseberg and Burke (2001), and Stebbins (1950) represent a minority BSC tradition viewpoint among botanists, while Wheeler (1999) represents a minority non-BSC viewpoint among zoologists. This striking distinction is probably mostly due to actual differences in reproductive biology among different branches of the tree of life. Specialists on organisms with either very open mating systems or highly clonal reproduction have always had trouble applying the BSC and have looked for alternatives.

The Hennigian phylogenetics revolution that began in the 1970s altered many aspects of theory and practice in systematics, but did not do much to prune the existing variety of species concepts, and in fact added several more. Hennig himself (1966) held to a version of the biological species concept, while other Hennigians preferred the *evolutionary species concept* (basically an interbreeding group viewed through time as a lineage; Wiley, 1978) or various versions of a *phylogenetic species concept*. The latter are a heterogeneous set of concepts as well: some quite similar to the phenetic species concept (i.e., species viewed as a unique set of character states; Cracraft, 1997; Nixon & Wheeler, 1990; Platnick & Wheeler, 2000; Wheeler & Platnick, 2000a, 2000b), others applying Hennigian concepts of apomorphy and monophyly to the species level (Mishler & Donoghue, 1982; Mishler & Theriot, 2000a, 2000b, 2000c; Rosen, 1978).

In comparing different views of species it is important to distinguish two components of any species concept: *grouping* vs. *ranking* (Horvath, 1997). The grouping component of any species concept indicates the criteria for group inclusion, whether ability to interbreed, phenetic similarity, or sharing of apomorphies indicating monophyly. The ranking component of any species concept indicates the criteria for deciding whether a group counts as a species rather than a taxon at some other rank. Both components are necessary because all concepts define groups within groups, and the level of group corresponding to species needs to be specified. Some of the controversy over species concepts has been because people are not clear about this distinction.

The phylogenetic species concept in the sense of my work with Brandon and Theriot (Mishler & Brandon, 1987; Mishler & Theriot, 2000a, 2000b, 2000c) is clear about this distinction, and basically treats species as just another taxon (see also Nelson, 1989), taking the perspective that if we are going to be phylogenetic about taxa in general, we need to be phylogenetic about species. Theriot and I (Mishler & Theriot, 2000a) defined species as follows: "A species is the least inclusive taxon recognized in a formal phylogenetic classification" (p. 46). As with all hierarchical levels of taxa in such a classification, organisms are *grouped* into species because of evidence of monophyly. Taxa are *ranked* as species rather than at some higher level because they are the smallest monophyletic groups deemed worthy of formal recognition, due to the amount of support for their monophyly and/or their importance in biological processes operating on the lineage in question. One obvious question follows from the definition given above: doesn't the ranking decision sound arbitrary? The short answer is: Yes! If not completely arbitrary, the decision does depend on local context – ranking criteria are pluralistic rather than universal (Mishler & Donoghue, 1982).

The ranking decision in the phylogenetic species concepts discussed above is present because of the way the current codes of nomenclature are written. Monophyletic taxa not only have to be discovered and diagnosed, they must be given a specific rank, including species. But this doesn't have to be so. We can remove this arbitrary aspect of taxonomy; the best approach is arguably not to designate any ranks at all. I now advocate an extension of the recent calls for rank-free phylogenetic taxonomy to the species level (e.g., Mishler, 1999; Pleijel, 1999), and will develop this position in the following sections.

Return to a Darwinian View of Species

Let's consider the two-part question introduced above: (1) *Are species real?* (2) *Are species uniquely real?* All working biologists today think that the answer to the first question is yes: species are real entities in some sense (although the grouping criterion considered to be the basis for their reality varies as described above). The current debate concentrates on the second question: whether or not species are a special level either in biological organization or in the taxonomic hierarchy. In other words, is there a unique ranking criterion for species? The two possible answers to this question can be contrasted as the *Darwinian view* vs. the *Mayrian view*.

One of Darwin's important novel contributions to biology was the explicit recognition that the species level is an arbitrary point in the divergence of two lineages. The *Origin* (Darwin, 1859) is full of passages indicating Darwin's view that the species rank is arbitrary, even though the lineages are quite real. His view was that divergence between two lineages happens, and at some point, it is convenient to call the two lineages species according to the judgment of a competent taxonomist, but nothing particularly special or universal occurs at that point.

The Modern Synthesis, in its bringing together of population genetics and taxonomy, emphasized a different point of view on species than Darwin. Species were now viewed as an important and distinct level of biological organization (like "molecule" or "cell"), the largest group within which evolution occurs, held together by sharing a gene pool. Ernst Mayr is particularly responsible for pushing this viewpoint (Mayr, 1982). Following Mayr, many today (scientists and public alike, in a strange convergence between evolutionary biologists and creationists) see species in this special light. Note that I am not calling Mayr or any evolutionary biologist a creationist. I am only pointing out an interesting parallel to their position in this one particular area. I don't think the parallel is an accident, however. I think that the idea of distinct, basic, natural units (i.e., species as the building blocks of biodiversity) is so ingrained in Western thought (coming from before the Christian era so not due to creationists directly) that most evolutionary biologists and ecologists have serious trouble letting go of it. Darwin was a really original and courageous thinker whom many biologists even today have trouble emulating.

There is abundant empirical evidence presented since Darwin's time that shows he had the right view and that the actual "species situation" is much more complex than modeled by the Modern Synthesis adherents (Mishler & Theriot, 2000a, 2000b, 2000c). Gene pools (potential horizontal transfer of genes at some level of probability) usually occur at many nested levels within one lineage, and the most inclusive level is often higher than anyone would want to call species (e.g., corresponding to the current generic and even familiar level in flowering plants). On the other hand, sometimes gene pools don't exist at all in a lineage, in the case of asexual organisms. Alan Templeton (1989) succinctly summarized this spectrum of problems with the Mayrian BSC as ranging from "too much sex" to "too little sex."

It would be conceptually cleaner if Mayr was right that there is a particular, unique level, comparable across the tree of life, at which "species-ness" arises as two lineages diverge. However, empirical reality intrudes on this tidy BSC concept; we need a more flexible concept since such a clean species break rarely if ever appears to be the case. My own view is that Darwin's richer conception is better, and that the supposed advances of the Modern Synthesis were actually retrograde, at least as far as species concepts are concerned. To make progress in this area we need to reject the simplistic Mayrian view and emulate Darwin's view.

My own answer on the twin questions italicized at the beginning of this section is this: entities that are currently called species are indeed real, if grouped correctly as monophyletic groups, but they are not uniquely real, i.e., they are only real in the sense that other levels of monophyletic groups are – there is no special ranking criterion for species. The processes causing divergence of lineages, and keeping them separate afterward, are many. We must develop a richer view of the tree of life and how best to understand and classify it. Such a view must consider the many nested levels of divergence and reticulation in the tree of life, not just the one we arbitrarily happen to call species.

To develop this view, we need to look closely at several related concepts. One is the nature of *monophyly*. There have been two basically different ways of defining monophyly within the Hennigian tradition of phylogenetic systematics: one is synchronic (i.e., "all and only descendants of a common ancestor"); another is diachronic (i.e., "an ancestor and all of its descendants"). I have argued elsewhere (Mishler, 1999) that the former view (Hennig's own view) is better, because it avoids the time paradoxes inherent in placing the ancestor in a group with its descendants. Just like a zygote is not one of the cells of an adult organism (instead it is *all* the organism at its beginning), the ancestor is not a member of a synchronic monophyletic group when looked at later – it was the whole monophyletic group back in its day.

A further consideration is that the word "species" appears in many definitions of monophyly (including Hennig's). This obviously matters if we are discussing the application of monophyly to the species level, because of circularity concerns. We need a definition that is both synchronic and neutral about taxonomic ranks, like this: a monophyletic group is all and only descendants of a common ancestor, where "ancestor" is interpreted broadly to mean an individual in the philosophical sense of Ghiselin (1974) and Hull (1978), e.g., an organism, or breeding groups of various sizes.

Another distinction that is needed is between *clades* and *lineages*. While sometimes treated loosely as synonyms, they are not exactly the same thing – some refinement of terminology is needed. Fig. 4.2 shows the difference. A "clade" is a synchronic entity, a monophyletic group as discussed above (a group composed of all descendants of a common ancestor). A "lineage," by

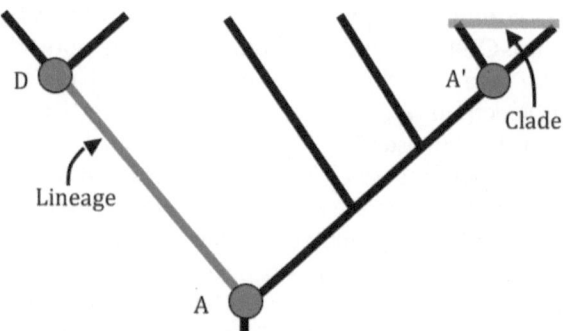

FIGURE 4.2 The distinction between clades and lineages. A clade is a synchronic, monophyletic set of lineage-representatives, where monophyly is defined synchronically as "all and only descendants of a common ancestor" (represented by A′ in this case). A lineage is a diachronic ancestor-descendant connection (between A and D in this case): "species" in the de Queiroz sense.

contrast, is a diachronic concept, a series of ancestors and descendants (replicators in the abstract sense of Hull, 1978) through time. They are related terms, of course: a clade could best be viewed as an instantaneous snapshot of a lineage.

This distinction helps us clarify some of the murky debates over phylogenetic species concepts. Some phylogeneticists have focused their species concepts on clades (e.g., Baum & Shaw, 1995; Mishler & Brandon, 1987; Mishler & Theriot, 2000a, 2000b, 2000c), and some on lineages (e.g., de Queiroz, 1999; Wiley, 1978), but it is important to note that both clades and lineages form hierarchies. Clades are obviously nested inside of other clades, but less widely understood is that the same is true of lineages. A smaller-scale lineage (say of cells) is nested inside of larger-scale lineages (such as organisms or larger groupings acting as individuals in a philosophical sense). There is no privileged level that can be recognized in either of these nested hierarchies; there is no unique species rank in either clades or lineages.

I prefer applying our formal classification system to name clades (i.e., monophyletic groups) for the following reasons: (1) clades are more nicely nested hierarchically than lineages; (2) we have a well-thought-out code of nomenclature available for naming clades (the PhyloCode); and (3) there are many more kinds of lineages, biologically speaking, due to the multiple kinds of replication which can occur in nature. Some recent workers have thought about providing a code to name lineages, i.e., a "Species Code" (see discussion in the PhyloCode preface at http://www.ohiou.edu/PhyloCode/preface.html), perhaps to complement the PhyloCode (which is based on clades), but this will prove to be very challenging.

The generalized view presented above, abandoning species in favor of describing clades at several nested levels, has many advantages in theory. Yet it requires considerable further explication before being applied in practice, since so many other areas of biology are accustomed to using species as a unit. I will go through several of these areas below, beginning with systematics, where it all starts.

Practical Implications

Truly Rank-Free Classification, All the Way Down

As covered in many previous papers (summarized by Mishler, 2009), it has become clear that the ranks in the Linnaean system are problematic for classification, both theoretically and practically. Let me just summarize these general arguments briefly here. Rapid advances in phylogenetic research have made it obvious that there are not nearly enough ranks to suffice in classifying the tree of life, with its thousands of nested levels of clades. The need to maintain the hierarchy of the ranks leads to instability, with names being changed without good reason, as, for example, when one currently recognized genus is found to be nested inside another (a common occurrence). Ranked classifications can lead to bad science in such fields as ecology or macroevolution, if a user of a classification naïvely (but understandably) assumes that taxa placed at the same rank must be comparable in some way.

The current codes of nomenclature can be tweaked to name monophyletic groups, but they are far from ideal for that purpose. The current codes are used to name *all* kinds of groups; thus a user has no way of easily knowing if a given taxon is thought to be monophyletic. Only a search into the literature can uncover the basis for a particular taxon name under the current codes, while under the PhyloCode one knows that the author of the name hypothesized it to be a monophyletic group. A name that can be used to convey anything really conveys nothing.

It has become clear that the current codes don't lend themselves well to naming monophyletic groups unequivocally, primarily because there is only one type specimen. It is possible to patch the current codes of nomenclature to name phylogenetic taxa, as suggested by (Barkley et al., 2004). But, for many reasons, it would be better to develop a new code of nomenclature specifically designed for phylogenetics. It really is time to bite the bullet and complete a synthesis between the Darwinian revolution and the Hennigian revolution (de Queiroz, 1988). Ranked classifications are a hold-over from the pre-Darwinian creationist mindset (Ereshefsky, 2002). They are not just a quaint anachronism; they are resulting in miscommunication at many levels. Completely rank-free phylogenetic classifications are far better for teaching, research, communicating with other scientists, and interfacing with the larger society.

What about the fundamental taxonomic level, species? Most published discussions about rank-free taxonomy are based on considerations of higher taxa alone, yet all the criticisms of taxonomic ranks summarized above can be extended to species – it is clear that all the arguments about the inadequacy of the current codes for naming phylogenetic taxa apply to the species level also.

The developing PhyloCode may be accessed online (http://www.ohiou.edu/PhyloCode/). This code maintains many of the features of current rank-based codes, but removes all ranks from clade names, and also uses multiple types (called "specifiers") to precisely fix the name of a clade. Important to this discussion, the current draft of the PhyloCode, unfortunately, does not deal well with providing names for what have been called species. Many uncomfortable special conventions are currently suggested for dealing with this particular rank.

Thus even the community of supporters of the PhyloCode is conflicted about what to do about species! More work is needed to make the PhyloCode work seamlessly at all taxonomic levels.

How could rank-free classification be applied to terminal taxa? Exactly as at other levels: names of all clades (including the terminal level) should be hierarchically nested uninomials regarded as proper names (current usage should be followed as much as possible to retain links to the literature and collections). As at all taxonomic levels, we could use either node-based or stem-based names with multiple internal specifiers (I personally think the use of apomorphy-based names is incoherent at any level, but that is another argument!). Specifiers should be actual specimens (this should be true at all levels).

In my opinion, species names should be converted from the current epithets (despite the current prohibition of this is the PhyloCode draft; see example of this in Fisher, 2006). The overriding principle is to achieve maximum continuity with previous literature for the sake of preserving connections to databases, literature, museum specimens, etc. There are two additional important principles, in my opinion: the naming system should be consistent for clades at all levels, and the PhyloCode should be distinct from the existing codes in terms of rules. In this approach, then, each clade named under the PhyloCode, including the terminal-most clade, has a uninomial given name, but also has associated with it a set of more and more inclusive "family names" (its clade address). In a database at least, all the higher clades to which a taxon belongs would be regarded as part of its complete name; this would help computers (and users) keep track of information in the database. Homonyms, which would result when converting species epithets to uninomials, can thus be told apart by higher-level clade names if their context is unclear, just as a teacher uses last names to distinguish among several children in class having the same first name.

Phylogenetic Monography

How can monographs be done under this view of species? In a rank-free framework, they can be done as well as or better under the current codes, as exemplified through the pathbreaking approach by Fisher (2006). Her approach was as follows: (1) use earlier taxonomies as a criterion for stratified-random selection of specimens to study (Hennig's semaphoronts); (2) after that, ignore taxonomic designation during character analysis and character scoring; (3) once operational taxonomic units are established (based on scored characters), conduct phylogenetic analysis; (4) use the resulting phylogenetic tree to inform taxonomic decisions, including naming of terminal clades consistent with the PhyloCode's treatment of more inclusive monophyletic groups. Specifiers used are specimens on deposit in an herbarium or museum, and the formal specifiers, as well as other specimens studied, are cited much as in traditional monographs. [See Fig. 4.1.]

DNA Barcoding

This discussion touches upon the current debates over DNA barcoding, another recently proposed system for characterizing species, which uses a short stretch of DNA sequence from a standard gene. Similarity above a certain percentage, say 2%, equals species status. This approach has gained popular appeal, but suffers from obvious philosophical problems. Contrary to their posturing as cutting-edge, DNA barcoders are actually returning to an ancient, typological, single-character approach, and are maintaining a pre-Darwinian view of species. There are two aspects to DNA barcoding, one good (but not new), the other new (but not good): DNA-based identification (i.e., using sequence data from a standard gene) and DNA taxonomy (i.e., using sequence data from a short stretch of a standard gene to recognize and name taxa). All critics (including me) are strongly in favor of the good idea of using DNA for identification of already well-characterized taxa, but that is old hat – the important use of DNA for identification goes back to the beginning of molecular systematics. The DNA barcoders can't take any credit for that – the most that they can claim is that they will scale-up, standardize, and database. But, there is really no need to set up a new bureaucracy or new databases (wasting the money of naïve funding agencies, who could be directing their attention toward real phylogenetic systematics) – current efforts elsewhere (such as GenBank) are more than sufficient. The new idea that DNA barcoding can replace normal taxonomy for naming new species and studying their relationships is not only bad philosophically, it is destructive in a practical sense. We should use all available resources to build real capacity to do systematics right (Will, Mishler, & Wheeler, 2005).

Implications for Ecology, Population Genetics, and Evolution

The species level is highly embedded in current ecological theory and practice. It is widely accepted that within- and between-species interactions are different in kind. Niche theory is usually conjoined with a view that the species level provides a fundamental break. Gause's (1934) theory of competitive exclusion talks about the ability of species needing to differentiate in order to live in the same environment. The species level is likewise highly embedded in studies of population genetics. The species is thought to be the largest unit in which gene flow is possible, thus the largest group that can actually evolve as a unit.

It is beyond the scope of this paper to elaborate on the ways to modify ecological theory to fit with a rank-free view of phylogenetic diversity (i.e., no species or other ranks). It needs to be done, however – based on the arguments presented above it is clear that the world is more complex than the current BSC allows for. If the systematic community moves to a rank-free view of biodiversity, then basic ecological and evolutionary research must be modified to account for this view. Fortunately, phylogenetic comparative methods are under active development in many areas (beginning with seminal studies such as Burt, 1989; Cheverud & Dow, 1985; Felsenstein, 1985; Harvey & Pagel, 1991; Huey & Bennett, 1987; and Martins, 1996). Studies can go forward on niche differentiation, competition, coexistence, species-area curves, community assembly, gene flow, macroevolutionary diversification, etc., but in a more rational manner taking into account nested hierarchical levels in these phenomena, without using ranks.

Implications for Conservation Biology

As argued in detail above, biodiversity isn't species – biodiversity is the whole tree of life, not just the arbitrary place at which species are named. There are clades smaller and larger than the traditional species level. Species are not comparable between lineages in any manner, just an arbitrary cut-off somewhere along a branch in the tree of life. Thus only someone sharing the BSC view that species are fundamental (a view interestingly shared by creationists, as discussed above) should think that species are the basic units of biodiversity, or that a list of currently named species in some way provides an inventory of biodiversity. Biodiversity is a much richer tapestry of lineages and clades.

So how can we inventory biodiversity without species? Since counting species or measuring their ranges and abundances is a poor measure of biodiversity, what should be done? New quantitative measures for phylogenetic biodiversity need to be applied which take into account the number of branch points (and possibly branch lengths) that separate two lineages. Phylogenetic measures of biodiversity have been developed that could be used as a basis for rank-free measures of biodiversity (Faith, 1992a, 1992b; Mishler, 1995; Vane-Wright, Humphries, & Williams, 1991). There are two possible approaches: counting of number of nodes separating two terminal clades, or summing the branch lengths separating two or more terminal clades. Advantages and disadvantages of each exist, and more work needs to be done, but the direction to move is clear.

What does "rarity" mean without ranks? This relatively new phylogenetic worldview can clarify greatly this term (Mishler, 2004). Rarity fundamentally means having few living close relatives, and these days "few" and "close" can be defined quantitatively on cladograms. Conservation priorities can actually be better guided by phylogenies rather than by taxonomy *per se*. Phylogenies provide a richer view of our knowledge of nested clades, and are directly associated with the evidence used to build them. Just like in the more theoretical areas discussed above, the most practical application of systematics in the modern world, conservation, needs to drop its reliance on species.

Postscript: Counterpoint

I agree with the quote at the beginning of Dr. Claridge's paper (about mountains in Switzerland) and with his statement that "species taxa represent attempts to recognize real biological entities." I believe mountains and taxa are real; as I explained in detail above, the issue for me is not whether taxa are real (they are, if monophyletic), but whether entities given the rank of species are real in a unique and special way that entities larger and smaller than them are not. Claridge and I agree that the entities we call species are real biological units. Our main difference is in what that reality is due to: for me it is monophyly, for Claridge it is sharing reproductive bonds. In either case, my point is that there are such real entities deeply nested inside each other, with no one level fundamental or unique. Species are real, but not in a unique and special way.

Claridge understates the fundamental differences between interbreeding groups and monophyletic groups; they are not the same thing theoretically or practically. In fact, they are diametrically opposed, by definition. As was first pointed out by Rosen (1978), the ability to interbreed is certainly a plesiomorphy and thus not a good guide to monophyly. It is the derived *inability* to interbreed, say the origin of a new mate recognition system, that can be an apomorphic feature useful to diagnose a monophyletic group. Any empirical test for reproductive compatibility is certain to be measuring plesiomorphic similarity. The BSC is (and should be) anathema to a cladist, which makes it puzzling how someone could be a solid cladist at all levels but species.

I agree with Claridge that breeding relationships are very complex and diverse – but would point out that this observation actually strengthens my point. There are smaller inbreeding groups (sometimes actual, sometimes potential) nested inside of larger interbreeding groups: local populations, clusters of populations, geographic regions, even up to the intergeneric level in flowering plants like orchids. The potential to successfully interbreed gradually trails off as one looks at more and more distantly related populations (as Darwin pointed out). Claridge acknowledges this when he says that "the process of speciation is a continuous one, so that drawing real lines between species as they evolve will be very difficult and intermediate stages must be expected," but then he contradicts himself in the same paragraph by saying: "species are of unique and real biological significance." In most organisms, there is no magical level at which the probability of successfully interbreeding suddenly drops from near 100% to near 0%. Thus even under the biological species concept, there is no unique and special level. Again, keep in mind the important distinction between grouping and ranking: breeding *groups* are very real – no one is denying that – it is the *ranking* decision about which level among many levels of nested breeding groups is to be called species that is arbitrary. Darwin was very aware of this distinction; we should still take his views seriously.

Evolutionary biology will be richer and much more accurate in its models of the world if this Darwinian hierarchical perspective is accepted. Evolutionary and ecological processes are occurring at many nested levels. "Speciation" is a major field of study, with many books and papers to its credit, which my point of view would seem to denigrate. But while I do think that "speciation" is an oversimplified concept, like the biological species concept on which it is based, I believe that there are important processes being studied by these researchers. I call it "diversification," the splitting of lineages influenced by a variety of interesting processes (ecological, reproductive, genetic, developmental, etc.). The important distinction I make is that diversification happens at many nested levels, not a single magical one, and full accounting of these is needed for a complete understanding of evolution. Focusing at the level of the entities taxonomists happen to call species in a particular case, as in standard studies of "speciation," is a one-dimensional look at a multidimensional, hierarchically nested process.

We can do better with a completely rank-free view of taxonomy. Claridge thinks that my discussion of rank-free classification is peripheral to our argument over species, but of course it is central to my position. The arguments against comparability of entities at a particular rank apply to "species" as much as "families" or "orders." Evolutionary processes are not just operating to produce what we happen to call species; they operate at many nested levels in producing the tree of life, "which fills with its dead and broken branches the crust of the earth, and covers the surface with its ever branching and beautiful ramifications" (Darwin, 1859, pp. 170–171).

REFERENCES

Barkley, T., DePriest, P., Funk, V., Kiger, R., Kress, W., & Moore, G. (2004). Linnaean nomenclature in the 21st century: A report from a workshop on integrating traditional nomenclature and phylogenetic classification. *Taxon, 53*, 153–158.

Baum, D.A., & Shaw, K.L. (1995). Genealogical perspectives on the species problem. In P. Hoch & A. Stevenson (Eds.), *Experimental and molecular approaches to plant biosystematics: Monographs in systematics, Volume 53* (pp. 289–303). St. Louis, MO: Missouri Botanical Gardens.

Burt, A. (1989). Comparative methods using phylogenetically independent contrasts. *Oxford Surveys in Evolutionary Biology, 6*, 33–53.

Cheverud, J., & Dow, M. (1985). An autocorrelation analysis of the effect of lineal fission on genetic variation among social groups. *American Journal of Physical Anthropology, 67*, 113–121.

Coyne, J., Orr, H., & Futuyma, D. (1988). Do we need a new species concept? *Systematic Zoology, 37*, 190–200.

Cracraft, J. (1997). Species concepts and speciation analysis – An ornithological viewpoint. In M. Claridge, H. Dawah, & M. Wilson (Eds.), *Species: The units of biodiversity* (pp. 325–339). London: Chapman & Hall.

Darwin, C. (1859). *On the origin of species by means of natural selection*. London: John Murray.

de Queiroz, K. (1988). Systematics and the Darwinian revolution. *Philosophy of Science, 55*, 238–259.

de Queiroz, K. (1999). The general lineage concept of species and the defining properties of the species category. In R. Wilson (Ed.), *Species: New interdisciplinary essays* (pp. 49–89). Cambridge, MA: MIT Press.

Ereshefsky, M. (2002). Linnaean ranks: Vestiges of a bygone era. *Philosophy of Science, 69*, S305–S315.

Faith, D. (1992a). Conservation evaluation and phylogenetic diversity. *Biological Conservation, 61*, 1–10.

Faith, D. (1992b). Systematics and conservation: On predicting the feature diversity of subsets of taxa. *Cladistics, 8*, 361–373.

Felsenstein, J. (1985). Phylogenies and the comparative method. *American Naturalist, 125*, 1–15.

Fisher, K. (2006). Monography and the PhyloCode: An example from the moss clade *Leucophanella*. *Systematic Botany, 31*, 13–30.

Gause, G.F. (1934). *The struggle for existence*. Baltimore: Williams & Wilkins.

Ghiselin, M. (1974). A radical solution to the species problem. *Systematic Zoology, 23*, 536–544.

Grant, V. (1981). *Plant speciation*. New York: Columbia University Press.

Harvey, P., & Pagel, M. (1991). *The comparative method in evolutionary biology*. Oxford, UK: Oxford University Press.

Hennig, W. (1966). *Phylogenetic systematics*. Champaign/Urbana, IL: University of Illinois Press.

Huey, R., & Bennett, A. (1987). Phylogenetic studies of co-adaptation: Preferred temperature versus optimal performance temperatures of lizards. *Evolution, 41*, 1098–1115.

Horvath, C. (1997). Discussion: Phylogenetic species concept: Pluralism, monism, and history. *Biology and Philosophy, 12*, 225–232.

Hull, D. (1978). A matter of individuality. *Philosophy of Science, 45*, 335–360.

Levin, D. (1979). The nature of plant species. *Science, 204*, 381–384.

Martins, E. (1996). Phylogenies, spatial autoregression, and the comparative method: A computer simulation test. *Evolution, 50*, 1750–1765.

Mayr, E. (1942). *Systematics and The origin of species from the viewpoint of a zoologist.* New York: Columbia University Press.

Mayr, E. (1982). *The growth of biological thought.* Cambridge, MA: Harvard University Press.

Mishler, B. (1995). Plant systematics and conservation: Science and society. *Madroño, 42*, 103–113.

Mishler, B. (1999). Getting rid of species? In R. Wilson (Ed.), *Species: New interdisciplinary essays* (pp. 307–315). Cambridge, MA: MIT Press.

Mishler, B. (2004). The underlying nature of biodiversity and rarity under a phylogenetic worldview, in relation to conservation. In M. Brooks, S. Carothers, & T. LaBanca (Eds.), *The ecology and management of rare plants of northwestern California* (pp. 183). Berkeley: California Native Plant Society.

Mishler, B. (2009). Three centuries of paradigm changes in biological classification: Is the end in sight? *Taxon, 58*, 61–67.

Mishler, B., & Brandon, R. (1987). Individuality, pluralism, and the phylogenetic species concept. *Biology and Philosophy, 2*, 397–414.

Mishler, B., & Donoghue, M. (1982). Species concepts: A case for pluralism. *Systematic Zoology, 31*, 491–503.

Mishler, B., & Theriot, E. (2000a). The phylogenetic species concept (*sensu* Mishler and Theriot): Monophyly, apomorphy, and phylogenetic species concepts. In Q. Wheeler & R. Meier (Eds.), *Species concepts and phylogenetic theory: A debate* (pp. 44–54). New York: Columbia University Press.

Mishler, B., & Theriot, E. (2000b). A critique from the Mishler and Theriot phylogenetic species concept perspective: Monophyly, apomorphy, and phylogenetic species concepts. In Q. Wheeler & R. Meier (Eds.), *Species concepts and phylogenetic theory: A debate* (pp. 119–132). New York: Columbia University Press.

Mishler, B., & Theriot, E. (2000c). A defense of the phylogenetic species concept (*sensu* Mishler and Theriot): Monophyly, apomorphy, and phylogenetic species concepts. In Q. Wheeler & R. Meier (Eds.), *Species concepts and phylogenetic theory: A debate* (pp. 179–184). New York: Columbia University Press.

Nelson, G. (1989). Cladistics and evolutionary models. *Cladistics, 5*, 275–289.

Nixon, K., & Wheeler, Q. (1990). An amplification of the phylogenetic species concept. *Cladistics, 6*, 211–223.

Platnick, N., & Wheeler, W. (2000). A defense of the phylogenetic species concept (*sensu* Wheeler and Platnick). In Q. Wheeler & R. Meier (Eds.), *Species concepts and phylogenetic theory: A debate* (pp. 185–197). New York: Columbia University Press.

Pleijel, F. (1999) Phylogenetic taxonomy, a farewell to species, and a revision of Heteropodarke (Hesionidae, Polychaeta, Annelida). *Systematic Biology, 48*, 755–789.

Rieseberg, L., & Burke, J. (2001). The biological reality of species: Gene flow, selection, and collective evolution. *Taxon, 50*, 47–67.

Rosen, D. (1978). Vicariant patterns and historical explanation in biogeography. *Systematic Zoology, 27*, 159–188.

Sokal, R., & Crovello, T. (1970). The biological species concept: A critical evaluation. *The American Naturalist, 104*, 127–153.

Stebbins, G.L. (1950). *Variation and evolution in plants.* New York: Columbia University Press.

Templeton, A. (1989). The meaning of species and speciation: A genetic perspective. In D. Otte & J. Endler (Eds.), *Speciation and its consequences* (pp. 129–139). Sunderland, MA: Sinauer Associates.

Vane-Wright, R., Humphries, C., & Williams, P. (1991). What to protect? Systematics and the agony of choice. *Biological Conservation, 55*, 235–254.

Wheeler, Q. (1999). Why the phylogenetic species concept? Elementary. *Journal of Nematology, 31*, 134–141.

Wheeler, Q., & Platnick, N. (2000a). The phylogenetic species concept (*sensu* Wheeler and Platnick). In Q. Wheeler & R. Meier (Eds.), *Species concepts and phylogenetic theory: A debate* (pp. 55–69). New York: Columbia University Press.

Wheeler, Q., & Platnick, N. (2000b). A critique from the Wheeler and Platnick phylogenetic species concept perspective: Problems with alternative concepts of species. In Q. Wheeler & R. Meier (Eds.), *Species concepts and phylogenetic theory: A debate* (pp. 133–145). New York: Columbia University Press.

Wiley, E. (1978). The evolutionary species concept reconsidered. *Systematic Zoology, 27*, 17–26.

Will, K., Mishler, B., & Wheeler, Q. (2005). The perils of DNA barcoding and the need for integrative taxonomy. *Systematic Biology, 54*, 844–851.

Species and Phylogenetic Nomenclature[19]

The motivation for the development of phylogenetic nomenclature (originally called "phylogenetic taxonomy") was to allow biological classification (or "systematization"), to represent phylogenetic relationships, and to embody important principles such as "the untenability of paraphyletic groups" (de Queiroz and Gauthier 1990). From this starting point, de Queiroz and Gauthier developed a creative new basis for systematization in which the entities are not ranked taxa but clades (de Queiroz and Gauthier 1990, 1992, 1994; de Queiroz 1992, 1994). Instead of attaching names to taxa by reference to a type and a rank, as in traditional biological nomenclature, phylogenetic nomenclature labels taxa by the use of multiple "specifiers" – specimens or apomorphic traits that unambiguously refer to a particular monophyletic taxon. The subsequent development of phylogenetic nomenclature included formation of a scholarly society, the International Society for Phylogenetic Nomenclature, and development of a formal Code, the PhyloCode (http://www.phylocode.org), that is now in final preparation for publication at University of California Press. The hope for many phylogenetic systematists was for a new system of nomenclature that would firmly connect taxonomy to phylogeny and would allow for more stable ways to name the many significant clades that make up the Tree of Life. However, a major challenge for the development of the PhyloCode has been the treatment of species. We argue in this paper that the approach taken to species in the PhyloCode is at odds with the motivations that drove many to support phylogenetic nomenclature and that these problems should be fixed before the PhyloCode is published and officially implemented.

The debate over the nature of species has plagued biological systematics throughout its history. Darwin (1859), in one of his most original contributions to biology, provided a cogent solution to the species problem, recognizing that species are merely labels applied to a continuous divergence process: during the divergence of lineages, there is no magical level that corresponds to the rank of species (Ereshefsky 2010; Mishler 2010). Under the Darwinian approach, particular species may be meaningful evolutionary groups of organisms, just like particular genera, sections, subspecies, and varieties, but the species rank is not qualitatively distinct from other ranks. The Modern Synthesis reversed Darwin's advance, attributing renewed significance to the species rank and presenting species as the unit of evolution (see Mishler 2010 for more discussion of this checkered history).

---

[19]N. Cellinese, D.A. Baum, and B.D. Mishler. 2012. Species and phylogenetic nomenclature. Systematic Biology 61: 885–891. [reprinted by permission]

Partly in reaction to the preeminent position given to species during the Modern Synthesis, the past few decades have seen the proliferation of species concepts (Mayden 1997, listed at least 24 different concepts). Although the development of these concepts has helped clarify alternative ways of prioritizing biological concepts and phenomena, there has certainly been no clear winner: We are no closer to finding a single universally accepted definition of species than we were in Darwin's time. Fortunately, the challenge of designing a system of phylogenetic nomenclature can (and should) be separated from the species concept debate.

The approach to species currently adopted in the PhyloCode was presented and defended by Dayrat et al. (2008). They stressed the incompatibility of Linnaean binomial nomenclature with phylogenetic nomenclature and concluded that species and clades are two different kinds of entities and that the former should not be accommodated within the PhyloCode system of rank-free nomenclature. In taking the stance that species are distinct from clades, Dayrat et al. (2008) adopted the particular conception of species proposed by de Queiroz (2007). According to this view, species are separately evolving but connected subpopulations (metapopulations), and as such species are not clades but rather segments of lineages. Although de Queiroz's lineage-based species concept has been criticized (e.g., Baum 2009; Hausdorf 2011; see also the critique below), the pros or cons of his species concept are not the most pressing issue here. Accepting that lineages are not the same entities as clades, and that the PhyloCode is about naming clades, our main disagreement with Dayrat et al. (2008) is that they still retain species in the PhyloCode. In doing so, they effectively impose on all would-be users of the PhyloCode their view that species are lineages. We believe, in contrast, that the only way to improve the PhyloCode, so that it can be used by anyone who wishes to name clades, regardless of what they think about the nature of species, is to remove all mention of species and to treat all clades, from the very smallest to the very largest, equally. This will allow systematists freedom to equate species with clades, or with lineages, or to deny the existence of species entirely.

de Queiroz's (2007) views on the nature of species have, unfortunately, (and unnecessarily) played a tremendous role in shaping the PhyloCode resulting in a Code that is not useable by the many individuals who are fully committed to the principles of phylogenetic nomenclature but do not accept this particular view of species. A system of nomenclature should not be tied to a particular philosophical outlook on something as controversial as the nature of species, especially a system focused on the naming of clades.

It is not our intent here to review the history of the debate over species and phylogeny or to anoint any particular species concept as the correct one. Instead, in this paper, we propose that the PhyloCode be amended to remove mention of "species" and to accommodate users who wish to be able to attach names to clades that approximate taxa at the traditional species level. Before doing so, we provide an overview of those principles that guide the system of phylogenetic nomenclature in general that are most relevant to the question of whether and how species names might be accommodated and then a brief history of how species have been treated in this system so far.

Background

Phylogenetic Systematics and Nomenclature

The motivation for phylogenetic nomenclature was to facilitate the precise naming of any monophyletic group in such a way that nomenclature can remain stable as long as knowledge of phylogenetic relationships remains stable. For the purposes of this discussion on species, it is important to clarify the concept of monophyly since "ancestral species" are mentioned in some definitions. The distinction between diachronic lineages and synchronic clades is illustrated in Fig. 4.3. Lineages are relationships through time between ancestors and descendants, whereas clades are composed of sets of tips that are in existence at any one time. A simple definition of monophyly, and hence clades, is: "a monophyletic group is all and only the descendants of a common ancestor" (Mishler 2010). The ancestor in this definition is not a species but rather a part of a lineage, such as an organism, kin group, or population (as discussed by Mishler and Brandon 1987). Also, as pointed out by Baum (2009), monophyly of a group does not (and can not) mean monophyly on every single gene tree – horizontal transfer and incomplete lineage sorting are frequent enough to ensure this. Instead, monophyly refers to an ensemble characteristic of organismic and/or genic descent.

Traditional nomenclature has limitations as applied to the modern understanding of the Tree of Life. First, the Tree of Life is a deeply nested hierarchy of monophyletic groups. There are far too many levels in this hierarchy for the

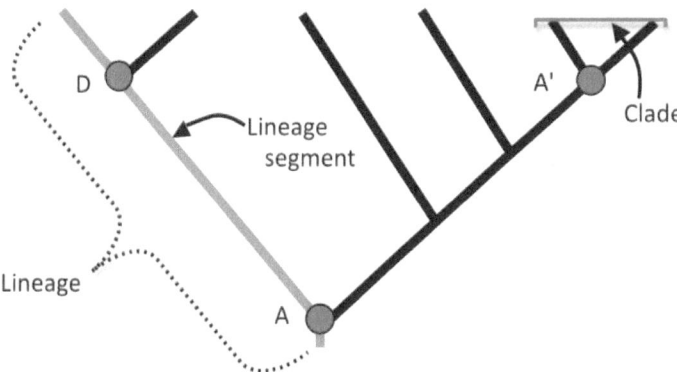

FIGURE 4.3   The distinction between clades and lineages, showing the incompatibility of different views of species. A clade is a synchronic, monophyletic set of lineage representatives, where monophyly is defined synchronically as "all and only descendants of a common ancestor" (represented by A' in this case). Note that the terminal-most named clade is a "species" in the sense of Mishler and Theriot (2000a, 2000b, 2000c). A lineage is a diachronic ancestor-descendant path (gray line up the left side of the tree), whereas a lineage segment is a part of a lineage that connects two nodes (A and D in this example), note that the latter are "species" in the de Queiroz (2007) sense. (Redrawn from Mishler (2010).)

number of ranks available in the traditional Codes of nomenclature. The rules in the traditional Codes that do not allow taxa to be nested inside taxa at the same rank lead to frustrating instability in names as our knowledge of phylogenetic relationships progresses. Second, formulating a taxon concept around a single type does not allow precise reference of a monophyletic group because it is not clear exactly what node on the tree is being named. As when triangulating in navigation, it is necessary to refer to at least two points of reference ("specifiers" in the PhyloCode) to unambiguously attach a name to a clade. Third, the traditional Codes are rank-based, but taxa at the same rank have nothing necessarily in common and are not comparable. Since ranking is highly subjective, whereas assignment of organisms to clades is largely objective (if not always easy), it seems prudent to build a nomenclatural system around clade affiliation rather than rank assignment. Clades exist regardless of rank, and the addition of rank to a clade, for example, whether they are referred to as genera or families, does not add any value to our nomenclatural system. For all these reasons and more, there has been a movement to develop a new code of rank-free phylogenetic nomenclature, the PhyloCode.

The development of a new nomenclature for species names has been almost as controversial as the species problem itself. A major point of contention between the traditional Codes and the PhyloCode is the lack of mandatory ranks in the latter. As a result, Linnaean species binomials are incompatible with phylogenetic taxonomy because they naturally imply the existence of both a genus rank and a species rank.

A number of initial proposals on how to treat species names in the light of phylogenetic taxonomy have been presented in the literature (de Queiroz and Gauthier 1992; Graybeal 1995; Schander and Thollesson 1995; Cantino 1998; Cantino et al. 1999; Ereshefsky 1999; Mishler 1999; Pleijel 1999; Pleijel and Rouse 2000, 2003; Artois 2001; Hillis et al. 2001; Dayrat et al. 2004; Dayrat 2005; Wolsan 2007; Baum 2009). These approaches worry primarily about how to convert traditional binomials into rank-free names that could be used in phylogenetic nomenclature. Sundberg and Pleijel (1994) proposed retaining the Linnaean species binomials for clades, with the traditional genus name serving as a prenomen: not part of the species name, but added by convention in order to retain historical name legacy. Cantino (1998) proposed, instead, the hyphenation of genus and epithet, thereby conserving the current generic name and including it as a part of the species name. However, this option was criticized because it would naturally lead people to consider species with the same prenomen as being closely related even though they might not be (Schander and Thollesson 1995). From that position, the combination of a higher taxon name (serving as a clade address) with a species uninomial would be preferred (Graybeal 1995; Schander and Thollesson 1995; Mishler 1999; Pleijel 1999). Although uninomials composed of species epithets have many virtues, they pose significant problems regarding their uniqueness within and across biological domains. A series of expedients to avoid name duplication were explored, using a combination of names followed by sequential numbers, registration numbers, etc. (see a full list of proposals in Cantino et al. 1999; Artois 2001; Wolsan 2007).

Dayrat et al. (2004) proposed to add three levels of uniqueness to species uninomials: (1) the traditional species epithet, (2) the original author's name, and (3) the date of publication. In cases where the three levels would still not provide uniqueness, page numbers and letters (a, b) could be added to the name string. Through such measures, it was shown to be possible to develop a clade-based naming system that encompassed taxa at or below the traditionally recognized species rank (Dayrat and Gosliner 2005).

In addition to worrying about the form of species names, some authors considered ways to make it clear from the name itself whether a certain taxon is at the species rank or not. Graybeal (1995) suggested the use of different kinds of parentheses to indicate species uninomials. Pleijel and Rouse (2000) recommended the use of an initial upper case letter for clade names that do not refer to least-inclusive taxonomic units (LITUs), and lower case for uninomials that do refer to LITUs. However, no agreement has been achieved as to the form of species names, mainly because these nomenclatural issues tended to be confounded with philosophical positions on the nature of species, in particular, whether species are clades or some other kind of biological entity and, if they are clades, whether the species rank is special.

It is worth highlighting that, regardless of one's view of species ontology, in many empirical studies that sample a large number of specimens, the set of organisms that were traditionally considered to comprise a single species are often found to approximately correspond to a clade. The same is of course true for sets of organisms that have been traditionally ranked as genera, families, etc., and are often found to correspond to clades. This is one reason why Baum (2009) argued that the PhyloCode ought to at least allow systematists to name clades that happen to approximately correspond in composition to a traditional species.

A Brief History of the Phylocode with Reference to Species

The first original draft of the PhyloCode was sketched at a workshop held at Harvard University in 1998 and was modified and subsequently made public in April 2000 via the Internet at www.phylocode.org. This draft covered the general application of clade naming rules that revolved around the philosophical principles of phylogenetic nomenclature. The publication of this early draft generated a large body of literature, some of which was supportive (Bremer 2000; Brochu and Sumrall 2001; Ereshefsky 2001; Laurin 2001, 2005; Bertrand and Pleijel 2003; Pleijel and Rouse 2003; Lee 2005 and many others) and some critical (Lucas 1992; Lidén and Oxelman 1996; Lidén et al. 1997; Dominguez and Wheeler 1997; Moore 1998; Benton 2000; Nixon and Carpenter 2000; Nixon et al. 2003, among others).

In 2002, a second workshop held at Yale University focused more on species names and the best approach to handle their nomenclature. Much disagreement characterized the debate and the issue was left open for further discussion, although the clear majority felt that a separate Code covering species would ultimately be necessary.

The First International Phylogenetic Nomenclature Meeting was held in Paris in July 2004. Further discussion about species and their names took place among the over 70 attendees. Different approaches were proposed, including two by Dayrat and Clarke (ironically, both authors in Dayrat et al. 2008) who argued for a uni-nominal system within the PhyloCode. Under both proposals, species were treated differently to clades: the species names being defined using the formula "the species that includes specimen X," with the author clarifying the kind of species entity to which the name refers. It was then agreed that Dayrat and Clarke should form a committee with the specific charge of drafting a Code for species along these lines.

The attempt to develop a phylogenetic code of nomenclature for species faced major challenges and ultimately failed due to the difficulty of reconciling several important issues including: (1) the realization that species defined under the traditional Codes vary widely in concept, (2) difference between ICZN and ICBN in how they handle species, (3) the conviction that every species name established under the current Codes would have to be converted into a PhyloCode name, and (4) the realization that lineages are nested inside of each other in complicated patterns meaning that if species are to be defined as lineages, they will nest at several hierarchical levels. Given these issues, the committee proposed at the Second International Phylogenetic Nomenclature Meeting held at Yale University in 2006 that species names continue to be established via the traditional Codes (ICBN, ICZN, and ICNB) and that subspecific groups be viewed as kinds of species. This proposal was favored by a clear majority of the meeting attendees, but with several nays (including the two authors of this paper who attended the meeting). This approach was then adopted by the committee that oversees the PhyloCode as described positively by Dayrat et al. (2008) in their summary of the recommendations:

> On May 24, 2007, the Committee on Phylogenetic Nomenclature (CPN) [...] adopted a new article in the PhyloCode addressing the naming of species in the context of phylogenetic nomenclature. This vote [...] represents a major step in the development of the PhyloCode.

This statement was meant by its authors to portray the ultimate achievement of the PhyloCode, but we feel that this vote represents instead a conclusive defeat for phylogenetic nomenclature. After years of debates in the literature, and at meetings and workshops, the difficulty of dealing with naming species in a phylogenetic context culminated with a surrender of species names to the traditional Codes.

Several revisions to the PhyloCode were made based on Dayrat et al. (2008). Notably, in the current PhyloCode draft (version 4c), art. 21 states that

> This code does not govern the establishment or precedence of species names. To be considered available (ICZN) or validly published (ICBN and ICNB), a species name must satisfy the provisions of the appropriate rank-based code (e.g., ICNB, ICBN, and ICZN) This article describes how species names governed by the rank-based codes are to be interpreted and used under this code.

Although acknowledging the value of democratic decision making, we remain deeply disappointed that the difficulty of dealing with naming species in a phylogenetic context culminated with a surrender not just of species taxa (which we are happy to let go) but species names as well. We here wish to argue that this poor decision should be revisited and reversed.

The General Lineage Species Concept and the Problems it Poses for the Phylocode

The de Queiroz (2007) approach to species is implicitly enshrined in the current version of the PhyloCode, which clearly states that clades and species do not equate (Art. 1.1) and implies that species are lineages (Rec. 21.4C). Many different species concepts are currently applied across the different biological domains, and many scientists find some of these useful whether conceptually, practically, or both. Many practitioners believe that species are ranked or unranked clades; many do not. But this disagreement is not relevant to a code of nomenclature that is rank agnostic. Rather than trying to attain a unified species concept, a sort of biological holy grail, we may have to just accept and embrace the diversity of thoughts about species and agree that phylogenetic nomenclature should remain logically independent from the philosophical debate about species. It is not constructive to anoint an "official" species concept in the PhyloCode, which is only a nomenclatural tool for naming clades not a ground for reconciliation.

We understand de Queiroz's position on species as segments of lineages and we agree with him that these are not the same kind of entity as clades. Lineages are diachronic, whereas clades are synchronic. Many authors since Hennig have appreciated this distinction, although it is sometimes muddled. Our disagreement with de Queiroz boils down to these points: (1) many currently named species are certainly not lineages. We have evidence for only a small number of them being actual lineages (it is empirically difficult to demarcate lineages as compared to the well-worked out evidential procedures for finding clades); (2) even when we can detect lineages, it is clear that nested levels of lineages can be discovered and it is unclear how one privileged level should be recognized as species. In other words, the species-ranking problem is in no way solved by equating species with lineages (regardless of exactly how lineages are defined). However, this disagreement should be irrelevant to the PhyloCode. The PhyloCode governs only clades. Other important biological entities such as lineages, genes, trophic levels must be named in other ways preferably in such a way that they are clearly distinguished from clade names (even in cases where they have the same nomenclatural root).

The PhyloCode as currently constructed works under the assumption that species are not clades, yet paradoxically brings species into the Code by stipulating that species names are disallowed for clades (Dayrat et al. 2008). When a clade happens to approximate a traditional species in content, the PhyloCode mandates that clade be given a new name, distinct from the traditional species name. However, when a clade happens to approximate a traditional genus or family (or any higher rank), it strongly recommends retaining and converting the traditional name whenever possible. This constraint is incompatible with several reasonable positions one might take on the nature of species and on the goals of phylogenetic nomenclature.

Some reasonable biologists believe in the existence of clades and doubt that species are meaningful real entities that warrant inclusion in a formal taxonomic system. For these individuals, the current PhyloCode poses severe problems because it inhibits their ability to name clades at or below the traditional species level. Likewise, a biologist who accepts that species are distinct from clades but who believes that a certain traditional species or subspecies name is best viewed as referring to a clade cannot apply that name to that clade.

Even more so, the PhyloCode discriminates against individuals who believe that species correspond to clades of some (perhaps arbitrarily determined) rank. These individuals not only have trouble naming clades at or below the species level but, when they discover a clade that closely approximates a named species, they are forbidden from given that clade its logical name – the relevant species epithet.

The current PhyloCode requires systematists to refer to species and subspecific taxa whose names are regulated by the traditional codes. The Code allows for no other way to utilize such taxa. Furthermore, the PhyloCode also allows, in fact, encourages the use of species taxa as specifiers in phylogenetic clade definitions (Art. 11.7). This approach poses some critical problems:

1. Having to depend on traditional Codes implies that the PhyloCode is not freestanding. We feel that the PhyloCode ought to be legally independent of the existing Codes (although some similarities in spirit are common to all, of course). In other words, no rules in the PhyloCode should depend on rules in the traditional Codes, including for species names. However, the PhyloCode as it stands does depend on traditional Codes, and these are not controlled by proponents of phylogenetic nomenclature. For example, ICBN could amend their Code to say that species names are invalid if, when published, there was intent to treat the genus name as a "prenomen" under the PhyloCode.
2. The current Code calls on systematists to name new species under the traditional Codes even if they intend only to use the names as specifiers within phylogenetic nomenclature (Art. 21.4). We feel that it is intellectually dishonest to assign a genus name to a new species (as required under the traditional codes) when in fact the "genus" is believed to be an artificial concept and its use would be later dropped by the author of the name. Even more dramatic is the case where a new species does not fall under any of the traditionally described genera, implying that the author should first establish a new genus under the traditional Codes before creating a new species, also subjected to traditional rules.
3. Species established under the traditional Codes and clades named under the PhyloCode are different entities, hard to precisely compare because they are fundamentally defined by a single-type and multiple specifiers, respectively. Therefore, adopting traditional binomials to label clades and their specifiers generates natural inconsistencies between clades and what traditional names actually represent. This is a fundamental reason why specifiers under the PhyloCode should be museum specimens (physical reference objects) at all levels, never Linnaean binomials. Similarly, a nested taxon can be equally used as a specifier of a larger clade, as long as the reference taxon has been previously defined according to the rules of the PhyloCode with reference to specimens.

4. In many cases, biologists discover clades with finer phylogenetic structure within a currently terminal taxon that approximately compare to species as circumscribed based on one of the many species concepts. In these instances, it seems logical to retain a clade reference to the original species name and allow that it contains smaller subclades requiring new names. This contrasts with what happens when a species is split under the traditional Codes, where the species name must be transferred down to one of the newly discovered subclades (the one containing the type), resulting in serious difficulties keeping track of synonymy in databases. Therefore, treating names of current terminal taxa just like clade names at all levels has a decided benefit for stability and biodiversity informatics. Yet this is banned under the current PhyloCode by rules that disallow on from assigning any species name to a clade.

A Solution to Naming Species in Phylogenetic Nomenclature

We suggest that the rules governing names of clades at and around the traditional species level should be identical to the rules that apply at higher levels. As there are no ranks under the PhyloCode, it seems illogical to have an implicit rank of species. However, Dayrat et al. (2008) define a position that is laid out in the current version of the PhyloCode, which effectively applies special rules at the "species" level in the PhyloCode hierarchy. Such an implicit ranking exists due to articles in the PhyloCode that imply or overtly specify that clades at or below the level of species are not governed by the PhyloCode and may not be given clade names. For example, it is allowed that the specifiers of clade names be species rather than particular type specimens. Because it is wrong to use species as specifiers for clades that approximate to traditional species or sub-clades thereof, this approach implies that clades cease to exist or cease to be nameable at the approximate level of species.

Furthermore, Article 10.9 of the PhyloCode states: "A clade name may not be converted from a preexisting specific or infraspecific epithet (ICBN and ICNB) or from a name in the species group (ICZN)." This implies that if we discover a clade that happens to conform in content to a traditional species, and we have developed a suitable definition for that clade, we are nonetheless forbidden from using the traditional species name for that clade. Such a restriction does not apply to genus-approximating clades, showing that the PhyloCode applies different nomenclatural rules to clades whose content approximates species than to other hierarchical levels.

We believe that in order to maintain continuity with past literature, systematists should be allowed to recognize clades that roughly correspond in content to traditional species (or subspecies) and to use their former epithets as uninominal clade names. We certainly do not advocate mass conversion of all existing species names established under the traditional Codes to clade names, as many of these, in fact, do not even correspond to clades. Furthermore, just as for higher clades, only relevant species names should be converted into clades names at the discretion of expert systematists. Fisher (2006) provides an example of a monograph treating clades that approximate species using rigorously defined uninomial clade names that would be invalid under the current draft of the PhyloCode.

We are fully aware that many epithets are very common within and between domains. However, homonyms can be fully resolved by the addition of references to the name string, for example, clade name author, year, registration number, and/or clade address, following up the strategy articulated by Dayrat et al. (2004) but applying it to all levels instead of only species. It is already determined that valid clade names will be assigned registration numbers by RegNum (Cellinese 2012), an on-line database, currently in its final stage of development that stores and manages phylogenetic definitions. This mechanism will readily allow users to redeploy a former species (or subspecific) epithet to clade without leading to ambiguity due to homonyms. For example, the epithet *digitata* is common to 299 plant species (IPNI). However, *Adansonia digitata* L. could be converted to *Digitata* Linnaeus 1759, 2:1144 [New Author] (*Adansonia*) RegNum-number, whereas *Ceinfugosia digitata* Cav. would convert to *Digitata* Cavanilles 1787, 3:174 [New Author] (*Ceinfugosia*) RegNum-number. We believe that naming the clade (*Adansonia*) *Digitata* need not prevent individuals recognizing the traditional species *Adansonia digitata* L. Even if the composition of these two taxa were to diverge over time, for example, because the species *A. digitata* came to be divided into two while the clade continued to refer to the larger grouping, we are confident that systematists would be able to communicate without confusion. After all, several competing ideas on species limits may coexist in the current system, yet systematists are quite able to indicate the taxon of interest in a particular case using conventions such as sensu. Although our proposal would result in a name string containing a few more components than traditional species names, these measures would largely remain behind the scenes in databases, ensuring the required uniqueness of clade names, and providing a stable link to important legacy data.

Conclusions

We do not agree with the assessment of Dayrat et al. (2008) that the PhyloCode's development is complete. Much progress still needs to be made before it is published. In particular, we cannot support the adopted approach for "resolving" the species problem by tying the PhyloCode to Linnaean binomials named under the traditional Codes. This approach, if enforced, will result in many previous supporters of Phylogenetic Nomenclature refraining from using the PhyloCode. The PhyloCode needs to be extended to cover all clades, including those that approximate to or are within traditional species, and should be made philosophically inclusive so that it can be adopted by anyone regardless of their preferred species concept.

If monophyly provides the basis for classification, then there is no reason why a clade-naming system should not be extended to cover all clades even those that may have similar content to someone's concept and/or rank of species.

Therefore, it seems clear to us that phylogenetic nomenclature must be able to handle clades at all levels and, for reasons of clear communication and fair reference to the legacy literature, that one ought to be able to assign them the traditional names (or some derivative thereof).

The species problem does not need to be resolved for the purposes of finalizing the PhyloCode. No species concept (including the ones advocated by the authors of the present paper) should be enshrined in the PhyloCode. Since its purpose is to name clades, any mention of "species" in the PhyloCode should be removed. We need to streamline the PhyloCode to focus solely on rational rules for naming clades at any level including the traditional species level.

The authors of this paper have submitted the needed modifications to the PhyloCode in a proposal to the Committee of Phylogenetic Nomenclature (Appendix 1, Dryad doi:10.5061/dryad.sr2hd7md) and sincerely hope that they, or some equivalent set of changes, will be adopted before the official publication of the PhyloCode. We hope that this paper will stimulate discussion and result in a more open-minded Code that can be applied to all clades in the Tree of Life.

Supplementary Material

Supplementary material including an online-only appendix can be found in the Dryad data repository at https://datadryad.org/stash/dataset/doi:10.5061/dryad.sr2hd7md.

Funding

N.C. is very grateful for support from the National Science Foundation (DEB-0953677) for work that fostered these ideas and D.B. acknowledges the National Science Foundation (DEB-0949121).

ACKNOWLEDGMENTS

We would like to thank Torsten Eriksson for feedback on an early draft of the manuscript, and Dan Faith, Matt Haber, Dick Olmstead, and an anonymous reviewer for constructive criticisms and valuable suggestions that helped improve our paper.

REFERENCES

Artois T. 2001. Phylogenetic nomenclature: the end of binominal nomenclature? Belgian J. Zool. 131:87–89.

Baum D.A. 2009. Species as ranked taxa. Syst. Biol. 58:74–86.

Benton M.J. 2000. Stems, nodes, crown clades, and rank-free lists: is Linnaeus dead? Biol. Rev. 75:633–648.

Bertrand Y., Pleijel F. 2003. Nomenclature phylogénétique: une reponse. Bull. Soc. Fr. Syst. 29:25–28.

Bremer K. 2000. Phylogenetic nomenclature and the new ordinal system of the angiosperms. In: Nordenstam B., El Ghazaly G., Kassas M., editors. Plant systematics for the 21st century. London: Portland Press. p. 125–133.

Brochu C.A., Sumrall C.D. 2001. Phylogenetic nomenclature and paleontology. J. Paleontol. 75:754–757.

Cantino P.D. 1998. Binomials, hyphenated uninomials, and phylogenetic nomenclature. Taxon. 47:425–429.

Cantino P.D., Bryant H.N., de Queiroz K.D., Donoghue M.J., Eriksson T., Hillis D.M., Lee M.S.Y. 1999. Species names in phylogenetic nomenclature. Syst. Biol. 48:790–807.

Cellinese, N. 2012. http://www.phyloregnum.org and http://wiki.flmnh.ufl.edu/regnum.

Darwin C. 1859. On the origin of species. London: John Murray.

Dayrat B. 2005. Advantages of naming species under the PhyloCode: an example of how a new species of Discodorididae (Mollusca, Gastropoda, Euthyneura, Nudibranchia, Doridina) may be named. Mar. Biol. Res. 1:216–232.

Dayrat B., Gosliner T.M. 2005. Species names and metaphyly: a case study in Discodorididae (Mollusca, Gastropoda, Euthyneura, Nudibranchia, Doridina). Zool. Scr. 34:199–224.

Dayrat B., Schander C., Angielczyk K.D. 2004. Suggestions for a new species nomenclature. Taxon. 53:485–491.

Dayrat B., Cantino P.D., Clarke J.A., de Queiroz K. 2008. Species names in the PhyloCode: the approach adopted by the International Society for Phylogenetic Nomenclature. Syst. Biol. 57: 507–514.

de Queiroz K. 1992. Phylogenetic definitions and taxonomic philosophy. Biol. Philos. 7:295–313.

de Queiroz K. 1994. Replacement of an essentialistic perspective on taxonomic definitions as exemplified by the definition of "Mammalia". Syst. Biol. 43:497–510.

de Queiroz K. 2007. Species concepts and species delimitation. Syst. Biol. 56:879–886.

de Queiroz K., Gauthier J. 1990. Phylogeny as a central principle in taxonomy: phylogenetic definitions of taxon names. Syst. Zool. 39: 307–322.

de Queiroz K., Gauthier J. 1992. Phylogenetic taxonomy. Ann. Rev. Ecol. Syst. 23:449–480.

de Queiroz K., Gauthier J. 1994. Toward a phylogenetic system of biological nomenclature. Trends Ecol. Evol. 9:27–31.

Dominguez E., Wheeler Q.D. 1997. Forum – taxonomic stability is ignorance. Cladistics. 13:367–372.

Ereshefsky M. 1999. Species and the Linnaean hierarchy. In: Wilson R.A., editor. Species: new interdisciplinary essays. Cambridge (MA): MIT Press. p. 285–305.

Ereshefsky M. 2001. The poverty of the Linnaean hierarchy. A phylosohical study of biological taxonomy. Cambridge: Cambridge University Press.

Ereshefsky M. 2010. Darwin's solution to the species problem. Synthese. 175:405–425.

Fisher K.M. 2006. Rank-free monography: a practical example from the moss clade Leucophanella (Calymperaceae). Syst. Bot. 31:13–30.

Graybeal A. 1995. Naming species. Syst. Biol. 44:237–250.

Hausdorf B. 2011. Progress toward a general species concept. Evolution. 65:923–931.

Hillis D.M., Chamberlain D.A., Wilcox T.P., Chippindale P.T. 2001. A new species of subterranean blind salamander (Plethodontidae: Hemidactyliini: Eurycea: Typhlomolge) from Austin, Texas, and a systematic revision of central Texas paedomorphic salamanders. Herpetologica. 57:266–280.

Laurin M. 2001. L'utilisation de la taxonomie phylogénétique en paléontologie: avatages et inconvénients. Biosystema. 19:197–211.

Laurin M. 2005. Dites oui au PhyloCode. Bull. Soc. Fr. Syst. 34:25–31.

Lee M.S.Y. 2005. Choosing reference taxa in phylogenetic nomenclature. Zool. Scr. 34:329–331.

Lidén M., Oxelman B. 1996. Point of view: do we need "phylogenetic taxonomy"? Zool. Scr. 25:183–185.

Lidén M., Oxelman B., Backlund A., Andersson L., Bremer B., Eriksson R., Moberg R., Nordal I., Persson K., Thulin M., Zimmer B. 1997. Charlie is our darling. Taxon. 46:735–738.

Lucas S.G. 1992. Extinction and the definition of the class mammalia. Syst. Biol. 41:370–371.

Mayden R.L. 1997. A hierarchy of species concepts: the denouement in the saga of the species problem. In: Claridge M.F., Dawah H.A., Wilson M.R., editors. Species: the units of biodiversity. London: Chapman and Hall. p. 381–424.

Mishler B.D. 1999. Getting rid of species? In: Wilson R., editor. Species: new interdisciplinary essays. Cambridge (MA): MIT Press, p. 307–315.

Mishler B.D. 2010. Species are not uniquely real biological entities. In: Ayala F.J., Arp R., editors. Contemporary debates in philosophy of biology. Malden (MA): Wiley-Blackwell. p. 110–122.

Mishler B.D., Brandon R.N. 1987. Individuality, pluralism, and the phylogenetic species concept. Biol. Philos. 2:397–414.

Mishler B.D., Theriot E.C. 2000a. The phylogenetic species concept (sensu Mishler and Theriot): monophyly, apomorphy, and phylogenetic species concepts. In: Wheeler Q.D., Meier R., editors. Species concepts and phylogenetic theory. New York: Columbia University Press. p. 44–54.

Mishler B.D., Theriot E.C. 2000b. A critique from the Mishler and Theriot phylogenetic species concept perspective: monophyly, apomorphy, and phylogenetic species concepts. In: Wheeler Q.D., Meier R., editors. Species concepts and phylogenetic theory. New York: Columbia University Press. p. 119–132.

Mishler B.D., Theriot E.C. 2000c. A defense of the phylogenetic species concept (sensu Mishler and Theriot): monophyly, apomorphy, and phylogenetic species concepts. In: Wheeler Q.D., Meier R., editors. Species concepts and phylogenetic theory. New York: Columbia University Press. p. 179–184.

Moore G. 1998. A comparison of traditional and phylogenetic nomenclature. Taxon. 47:561–579.

Nixon K.C., Carpenter J.M. 2000. On the other "phylogenetic systematics". Cladistics. 16:298–318.

Nixon K.C., Carpenter J.M., Stevenson D.W. 2003. The PhyloCode is fatally flawed, and the Linnaean system can easily be fixed. Bot. Rev. 69:111–120.

Pleijel F. 1999. Phylogenetic taxonomy, a farewell to species, and a revision of *Heteropodarke* (Hesionidae, Polychaeta, Annelida). Syst. Biol. 48:755–789.

Pleijel F., Rouse G.W. 2000. Least-inclusive taxonomic unit: a new taxonomic concept for biology. Proc. Biol. Sci. 267:627–630.

Pleijel F., Rouse G.W. 2003. Ceci n'est pas une pipe: names, clades and phylogenetic nomenclature. J. Zool. Syst. Evol. Res. 41:162–174.

Schander C., Thollesson M. 1995. Phylogenetic taxonomy – some comments. Zool. Scr. 24:263–268.

Sundberg P.E.R., Pleijel F. 1994. Phylogenetic classification and the definition of taxon names. Zool. Scr. 23:19–25.

Wolsan M. 2007. Naming species in phylogenetic nomenclature. Syst. Biol. 56:1011–1021.

The Hunting of the SNaRC: A Snarky Solution to the Species Problem[20]

Abstract

We argue that the logical outcome of the cladistics revolution in biological systematics, and the move towards rankless phylogenetic classification of nested monophyletic groups as formalized in the PhyloCode, is to eliminate the species rank along with all the others and simply name clades. We propose that the lowest level of formally named clade be the SNaRC, the Smallest Named, and Registered Clade. The SNaRC is an epistemic level in the classification, not an ontic one. Naming stops at that level because there is no currently acceptable evidence for clades within it, not because no smaller clades exist. Later, included clades may be named. They would then become the SNaRCs, while the original SNaRC would keep its original name. We argue that all theoretical tasks of biology, in evolution and ecology, as well as practical tasks such as conservation assessment, are better approached using this rankless phylogenetic approach.

Keywords

species, species concept, monophyly, phenomena, explanation, taxonomy

"For the Snark's a peculiar creature, that won't be caught in a commonplace way".

–Lewis Carroll, *The Hunting of the Snark* (1876)

Introduction

Species are often thought to be the fundamental units of evolution, ecology, genetics, and/or systematics (Agapow, Bininda-Emonds, et al. 2004; Birky, Adams, et al. 2010; Blaxter, Mann, et al. 2005; Claridge, Dawah, et al. 1997; Green 2005; Hull 1975; Reydon 2005). However, species are empirically unstable objects, being revised regularly. Moreover, they are theoretically unstable as well; discussions over what counts as a species, and what criteria are to be used to delineate one from another, show no signs of abating (Hausdorf 2011; Naomi 2011; Staley 2013; Wilkins 2011). Some have suggested the other extreme: that species are no more fundamental than monophyletic taxa (that is, those having a single common ancestor) at any rank (Mishler 2010), and furthermore that the

[20]B.D. Mishler and J.S. Wilkins. 2018. The hunting of the SNaRC: a snarky solution to the species problem. Philosophy, Theory, and Practice in Biology. 10: 1–18. [reprinted by permission]

species rank should disappear as part of a general move to rankless taxonomy (Ereshefsky 1999; Mishler 1999; Pleijel 1999).

As it stands, most taxonomists resist rankless taxonomy and think that ranked taxa, particularly species, represent natural kinds or explanations of biological groupings. Some (Fitzhugh 2005, 2009) even think that delineating taxa such as species are explanations in themselves. Even many of those committed in general to the idea of rankless taxonomy make an exception for species and think that this one taxonomic rank is important to keep (Cantino and de Queiroz 2000; de Queiroz 2005, 2007). Other supporters of rankless taxonomy advocate a consistent treatment of names for all levels of clades under the PhyloCode, the revision of taxonomy proposed to replace the Linnaean scheme (see below). That includes even the levels corresponding to what have been called species (Cellinese, Baum, et al. 2012).

This topic is thus maximally controversial along several axes of opinion. Are species unique entities of biodiversity or are they the same as taxa at higher and lower levels, either within a ranked or rankless nomenclatorial system? If they are unique, what is their supposed uniqueness due to?

Background About Species

Since the inception of modern botany and zoology, biologists have had the notion of a "good species," and although it is not universally agreed what that means, each taxonomist has little trouble in identifying these entities in their chosen study group of organisms, through some sort of prototypic approach (Amitani 2015). Such prototypes have a folk taxonomic origin. All societies have folk concepts of living kinds of one kind or another (Medin and Atran 1999), albeit usually not in a formally ranked nomenclature. In most early classifications, taxa were not ranked. While folk kinds are generally nested in a hierarchy, there are no fixed and specified levels or grades of kinds in most folk taxonomies (Atran 1990, 1999), and prior to the early modern period in the life sciences ("natural history"), there was no such rank either (Wilkins 2018). Thus, species were named before there was any biological theory to speak of, first by botanists and later by zoologists. Christian theological considerations impinged on the question of rank early. The need for a species *rank* appears to have arisen as a result of attempts, by Johannes Buteo and Athanasius Kircher in the 16th century (cf. Breidbach and Ghiselin 2006; Buteo 1554; Kircher 1675; Wilkins 2013), to work out logistically how many kinds (*species* in Latin) were on the Ark. Other pre-Darwinian theorists of classification also viewed higher ranks such as phyla as indicating major elements of God's creative plan (Agassiz 1859). Ranked classifications, in general, are a late innovation, and one due to professional and historical contingencies in modern biology (McOuat 2001).

Only after Darwin did the question "What makes a species a species?" come into general discussion. It emerged particularly after Johanssen's failed "pure lines" argument during the Mendelian revolution, in which he held that species were pure gene lineages (Wilkins 2010, 2011). The species problem, as it came to be known, became a central issue of the Modern Synthesis with Dobzhansky

(1935, 1937). Prior to the Mendelian revolution, there was a "species question" (*What is the origin of species?*), addressed by Darwin, but not a "species problem" (*What are species?*) (Wilkins 2013b).[21] Darwin, who addressed the species question, clearly thought species were simply less transient varieties and left the species problem at that; evolution did not license ranking. Many subsequently concluded (erroneously) that Darwin thought were no such things as species, when what he really thought was that the species *rank* was arbitrary (Mishler 2010; Wilkins 2018).

Taxa are supposed, in any biological classification, to represent or name natural groupings, and *monophyly* is the modern criterion of a natural group in phylogenetics. This criterion, however, admits of more than one interpretation, in part depending on whether the monophyletic group is thought of as *synchronic* (representing a single time slice across lineages descended from a common ancestor) or *diachronic* (representing the historical causal relations between a common ancestor and its descendants). In other words, is the ancestor included in the group (diachronic) or not (synchronic)? The diachronic view that a group is monophyletic if it includes a single historical common ancestor we might call *dia*monophyly (Vanderlaan, Ebach, et al. 2013). The synchronic view that monophyly is a property of extant or extinct taxa in a simple relationship to each other, such that a group of specimens are more closely related to each other than to an outgroup, we might call *syn*monophyly.

However, what are commonly regarded as a "good species" are often not monophyletic under either definition. Named species can be formed by repeated speciation (Turner 2002) or by hybridization between other species (Bogart 2003; Rieseberg 1997). Or particular species may have an incomplete coalescence of genetic lineages or haplotypes (Beltrán, Jiggins, et al. 2002; Després, Pettex, et al. 2002).

Mayr made an important distinction between species as particular *taxa* and species as a general *category*. Mayr held that both a given species taxon and the species category were natural and therefore real (Mayr 1996). Others (known as species nominalists) have held that neither species taxa nor the species category are natural or real (Hey 2001, 2006; Pleijel 1999; Vrana and Wheeler 1992). Still, yet others (e.g., Darwin, Mishler) have held that species as individual taxa can be natural but that the species category (the rank) is unnatural. Let us first consider these three approaches to the species category, or rank, before offering a solution in terms of monophyly.

Approaches to Species as a Category

1. *Theoretically defined species:* Mayr formulated and promoted a theoretical account of the species category, the Biological Species Concept (BSC), which many biologists adopted uncritically. Widely held during the twentieth century, this is the view that a group of organisms is a species if and only if it satisfies a theoretical criterion,

---

[21] The observation is due to Jody Hey (pers. comm.). Contrary to Mayr's and others' characterizations, Darwin did not intend to define species, but to explain why they existed.

specifically that it is reproductively isolated. One major problem with a theoretically-defined species account such as this is that it excludes groups that are empirically regarded by biologists as "good species" but which do not satisfy its criterion. Examples include asexual organisms (Bogart 2003; Lodé 2013; Moritz and Bi 2011), mostly asexual (yet occasionally prolifically exchanging genes) microbial organisms (Ochman, Lerat, et al. 2005; Wilkins 2007a), and, as mentioned, hybridizers.

The BSC is not the only possible theoretical species concept, of course. For example, the Ecological Species Concept (Van Valen 1976) uses filling an ecological niche as a criterion, and the General Lineage Concept (de Queiroz 1999, 2007)[22] uses forming a lineage as a criterion. Both, along with the BSC, have been criticized for empirical vagueness of application and lack of congruence with what are empirically regarded as "good species."[23] Altogether, there are around twenty-eight theoretical definitions of the species category (Wilkins 2018). None match all and only the species taxa empirically identified by taxonomists and ecologists.

The well-known "species-as-individuals" thesis (the SAI or Individualist Thesis) was built on a philosophical foundation: individual species are unique historical objects, held together by theoretically important processes, rather than natural kinds (Gayon 1996; Ghiselin 1974; Hull 1976; cf. Wilkins 2007b). These important processes were most often considered to be reproductive cohesion, although other processes could be involved (see Ghiselin 1997; Mishler and Brandon 1987, for discussion). Under this view, organisms are *parts* of a species, rather than *members* of it, and species are seen as occupying a unique level in the tree of life. This stands in contrast to the philosophical view that there are "natural kinds," in which members of a species share a unique set of traits (Dupré 1981; Hacking 1990, 2007; Khalidi 2013; Rieppel 2010; Wilkins 2013a).

2. *Antirealism about species:* Another widespread view is that the species category and sometimes also particular species taxa are unreal objects in biology. On this view, they are merely conventional terms, names without name-bearers. This is often held erroneously to be the "Darwinian" view (Wilkins 2009, 129ff), although it really came into prominence around 1900 (Anon 1908). This view is also sometimes called "species nominalism," based on the medieval philosophical position of nominalism, which holds that only individual objects exist. This

---

[22] It is misleadingly also known as the "Unified Species Concept." This term is misleading because, although all species to the extent they are natural objects form lineages, that is also true of taxa at all ranks. It is entirely unclear what kind of lineages uniquely qualify to be named at the species rank. So this conception is unified (and general) just to the extent that it proposes a necessary but insufficient criterion for a natural species concept.

[23] "Good species" form the proof of concept for biologists for concepts of species (the rank). Every biologist knows what form good species take in their specialty, but each subdiscipline differs in subtle or gross ways from other subdisciplines. See Amitani (2015) for a discussion of this and a characterisation of "good species" as a form of prototypical reasoning.

view reached its apogee in the phenetics era in taxonomy (e.g., Levin 1979; Sokal and Crovello 1970), but it still has advocates today.[24]

3. *Monophyletic species:* Many cladists,[25] but by no means all, have taken an intermediate position, between the two previous views. They hold that individual species taxa can be real objects in biology if they are monophyletic, but also that the species category is neither natural nor uniquely real since there are monophyletic groups at many levels. This is sometimes called the *phylogenetic species concept* (PSC), although, as with monophyly, there are at least two contenders for the PSC label, one based on historical monophyly (Mishler and Theriot 2000), and one based on diagnostic characters (Wheeler and Platnick 2000). A major problem perceived by some for the PSC is that there is often monophyletic structure below "good species" that leads to potential *taxonomic inflation*, an explosion in the number of species that are described (Isaac, Mallet, et al. 2004; Zachos and Lovari 2013). Likewise, it has been argued by some that the use of a phylogenetic conception based on monophyly could lead to excessive lumping (Staley 2006). Both criticisms result from the fact that the use of monophyly for species needs some criteria external to monophyly to decide which "level" of monophyletic group to rank as species. These external criteria vary from group to group (Mishler and Donoghue 1982), and under this view, it is difficult to see species as a unique level in the tree of life (Mishler 1999, 2010).

A Way Forward: Species are at Least Initially Phenomena

It is perhaps feasible to adopt a somewhat different approach, in this as in other epistemic matters in science: to take species as *phenomena* to be accounted for.[26] This means treating species as *explananda* (things which need to be explained) rather than as *explanantia* (things which explain). As discussed above, folk taxonomies demonstrate that human cultures generally perceive species as phenomena. Phenomena have been deprecated in the philosophy of science since the theory-observation dichotomy was criticized and abandoned. Recently,

---

[24] Numerical taxonomy, also known as the phenetics school (from the Greek phaineros for "appearance") classified groups according to their "overall similarity." This fell prey to the problems discussed by Nelson Goodman; as he says, similarity is cheap (see Decock and Douven [2011] for a discussion): "Similarity, I submit, is insidious. And if the association here with invidious comparison is itself invidious, so much the better. Similarity, ever ready to solve philosophical problems and overcome obstacles, is a pretender, an impostor, a quack. It has, indeed, its place and its uses, but is more often found where it does not belong, professing powers it does not possess." (Goodman 1972, 437) Depending on the characters used, phenetic groups, known as Operational Taxonomic Units or OTUs, could contradict other analyses using different characters of the same organisms.

[25] Cladism is the approach to classification that defines taxa by uniquely shared common ancestry (monophyly), as evidenced by shared derived characters. It is also known as phylogenetic systematics.

[26] Existing species concepts (except the conventional ones) define species in terms of some model or process, which is to say, as entities of a particular theoretical kind. To treat species as phenomena in need of explanation is to not beg the question in favor of a prior mechanism, which we take to be a scientific virtue. Thanks to a reviewer for raising this question.

though, phenomena have been revisited as a source for scientific discovery (Apel, Dullstein, et al. 2009; Massimi 2008, 2011; Schindler 2011; Woodward 2000). Phenomena represent a relation between the observer/classifier and the world (Bogen and Woodward 1988; Schindler 2011). Observed phenomena can represent real states of the world, but they are dependent upon the pattern recognition capacities of the observer (Wilkins and Ebach 2013). They do not necessarily rely upon prior definitions in order be observed. Some observed phenomena may of course turn out *not* to represent real states of the world; astrological signs or astronomical constellations which are recognized in different cultures are in this category. Phenomena should be treated by science as things to be explained that can be dissolved upon further analysis, and which are often revised.

As an example of a real phenomenon, take a standard philosophical case: mountains. We can identify examples of mountains, but not universally define a category of "mountain" as distinct from "hill" or "plateau." There is no standard height, geological basis, or other property that defines all and only mountains, and what may be called a mountain in Australia, for example, is a mere hill on most other continents. There is no fundamental hierarchical level of "mountainhood" – within what is recognized as a major mountain range there are recognized sub-ranges and individual peaks. Therefore, in terms of the argument in this paper, "mountain" is a rankless concept. Nevertheless, the reality of even an Australian mountain can be demonstrated by the fact that to get to the other side of it, one must go over, around or through it. Though the *category* is a construct, that does not mean the *individual objects* delineated within the category are constructs.

Phenomenal Taxa

Phenomenal taxa are evident groupings of organisms at all levels that have been apparent to folk going back thousands of years. Their perception does not initially rely upon prior definitions – the criteria used are operational and rely on covariances of traits of all kinds, most often the so-called morphological kind.[27] Phenomenal taxa are patterns that call for hypotheses; defensible scientific classifications result when phenomenal taxa are tested, and sometimes dissolved or at least revised on the basis of subsequent assays and phylogenetic analyses.[28]

Phenomenal Species

As discussed above, the lowest level of phenomenal taxa can be recognized as phenomenal species, starting with folk classifications (Atran 1999).[29] Species

---

[27]Which traits are selected to use for such comparisons depend a lot on prior experience rather than theoretic criteria, in traditional societies as well as in modern taxonomy.

[28]See Scerri (2007) for an example from chemistry, the periodic table. The properties of elements were experimentally measured and the periodicity of these properties noted before any theoretical explanation (such as valency theory or electron shells and proton number) was available. Likewise, plate tectonics was observed as a phenomenal pattern before an explanation was offered (Oreskes and LeGrand 2003).

[29]There is an extensive literature on folk taxonomy. We simplify here and are not suggesting that the same basic taxa are recognized in all or even most cultures (Berlin 1973, 1976; Berlin, Breedlove, et al. 1973; Durkheim and Mauss 1963; Medin and Atran 1999; Sousa, Atran, et al. 2002; Zachar 2000).

phenomena set up the conditions for an explanation, and once these are offered we may revise them, or even dissolve them. Many commonly recognized species turn out not to represent natural states of the world, now that the science of systematics has a good theoretical understanding of monophyly and a good set of empirical tools. New tools have emerged from rapid technological advances in computer hardware and software and in molecular biology. On the other hand, many commonly recognized species have been confirmed this way.

There is an old distinction between an *explanandum* and an *explanans*, or roughly the phenomena and the theory. In these terms, species do not explain anything; instead, they set up the problem that theoretical explanations solve. Species are *explanantia*. It used to be said that theories "save" the phenomena; this came to mean that they would solve the problems posed by explaining the phenomena (Hacking 1983, 222f). The advantage of the phenomenal approach is that it is empirically driven but not absolutely ranked.

Rankless Taxonomy

As indicated briefly above, the systematics community has reached a consensus that monophyly is the best criterion for a *natural group*[30] in classification. This consensus comes from several important criteria of classification (Mishler 2009; Wilkins and Ebach 2013), including information content (summarizing what is known about organisms), predictivity (what is not yet known about organisms), and function in theories (capturing entities involved in important natural processes). The latter tracks causal relations (e.g., evolution from common ancestors) even in the absence of detailed knowledge of those causes.

Given the adoption of monophyly as the basis of classification, many empirical and theoretical problems have arisen with the current codes of nomenclature with their mandated taxonomic ranks. There are not enough ranks to name the thousands of levels of clades that have been discovered, and instability is introduced when groups at the "same" taxonomic rank are found to be nested inside of each other. Most importantly, it is impossible to precisely specify a named clade with only one type specimen (de Queiroz and Gauthier 1992, 1994). The idea of removing ranks from nomenclature has developed as a response to these perceived problems and has gained much support and much criticism.

The major attempt to develop a rankless code of nomenclature to date is the PhyloCode (Cantino and de Queiroz 2000). Its basic philosophical foundation is that all groups should be natural (monophyletic), rank-free, and uniquely identified. Though there are no ranks, there is still hierarchy, as names are nested within names. The PhyloCode uses two or more type specimens (called "specifiers") to triangulate precisely to the clade being named. All new names are registered in a database (REGNUM[31]) with associated metadata including specifiers; the hierarchical nesting of clade names is thus clear, and a name can be applied stably into the future.

---

[30] In taxonomy, a natural group is a real fact about the organisms' interrelationships. It is contrasted to artificial groups, which are merely conventional, or based upon arbitrary characters.

[31] http://www.phyloregnum.org/

However, even within the PhyloCode community, there is vehement controversy about species (see discussion in Cellinese, Baum, et al. 2012). The PhyloCode as it stands retains species as a special level and explicitly excludes the normal application of uninomial names to clades recognized at the traditional species level. This, despite the obvious paradox of a code designed for rankless classification retaining one rank as privileged! This issue is an area of active debate, and it is unclear how it will be resolved in the long run in the PhyloCode. Regardless, for the solution to the species problem proposed here, we argue in favor of the application of concepts of monophyly and rankless phylogenetic classification "all the way down."

Capturing the SNaRC

What *are* the natural objects of taxonomy? If monophyly is the criterion, then we must recognize and name the most differentiable clades based on the data available, and the assays used (that is, the differentia that we can access as characters and character states). These were mostly morphological traits prior to the molecular revolution, and subsequently mostly gene sequences, but the methodology remains the same. Taxa, including species, are first recognized as phenomena, then tested by phylogenetic analysis.

Following the principles of rankless taxonomy and the synchronic definition of monophyly, the smallest named clade should be treated like other levels and given a formal (uninomial[32]) name registered in a database. These we call the *Smallest Named and Registered Clades* (SNaRCs).[33] Note that this is the smallest level in a hierarchical classification of clades in an *epistemological* sense rather than an *ontological* one. These are the finest-scale clades that can be convincingly demonstrated with *current* data; no claim is made that they are the smallest clades that exist in that group. Further research in the next generation may well find clades within what was regarded as a SNaRC in the current generation (in short, new biodiversity has been discovered). In that case, the finer ones are now the SNaRCs but the original clade retains its name. Thus, the completely rankless naming system is much more stable than the current codes of nomenclature which use a binomial and one type specimen. Currently, if a species is split into finer species taxa, the binomial name must move to one of the finer taxa, causing endless problems with using species names to organize databases and match comparable data.

It is important to note, however, that the cladistic concept of *monophyly* is itself in need of refinement. Horizontal transfer (reticulation) is much more

---

[32] If monophyly is the key, then the name or structure of the including clade is not relevant to the identification of that smallest (currently known) clade. Hence a genus name, giving a binomial, is not necessary.

[33] This is similar in some respects to Pleijel and Rouse's notion of a Least Inclusive Taxonomic Unit, or LITU (Pleijel and Rouse 2000), in that there is no rank that is fundamental. They say "Identification of taxa as LITUs are statements about the current state of knowledge (or lack thereof) without implying that they have no internal nested structure." However, with SNaRCs we restrict the naming of terminal taxa to entities regarded as clades, by requiring the author of a SNaRC name to present evidence of monophyly.

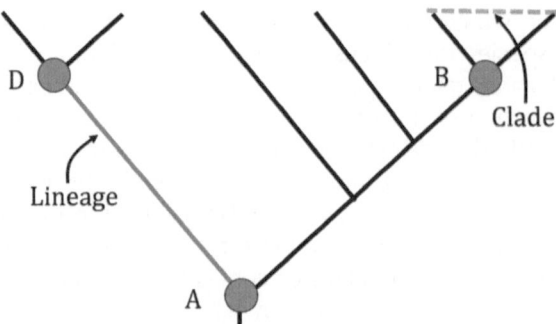

FIGURE 4.4 The distinction between clades and lineages. A clade is a synchronic, monophyletic set of lineage-representatives, where monophyly is defined as "all and only descendants of a common ancestor" (represented by *B* in this case). A lineage is a diachronic ancestor-descendant connection (between *A* and *D* in this case): "species" in the de Queiroz sense.
(Redrawn from Mishler 2010).

common in nature than realized twenty years ago (see a nice summary by Mallet, Besansky, et al. 2016). Despite having been presented as such (e.g., Wheeler and Platnick 2000), reticulation is not just a problem for the species level; clades at all levels can be subject to horizontal transfer.[34]

In the modern genomic world, because of the mounting evidence of horizontal gene transfer at all levels, monophyly can no longer mean monophyly of a group of organisms on every gene tree (as assumed by earlier generations of cladists, before there were data to the contrary). We would have few to no monophyletic groups, at any level, in that strict sense. Rather, monophyly refers to an ensemble characteristic of organismic descent as discussed by Baum (2009). Monophyly refers to the *preponderance* of gene lineages making up a clade (using the clade-lineage distinction from Mishler 2010; see Fig. 4.4). Gene lineages that don't match the pattern of descent shown by the majority of lineages need a different explanation (e.g., horizontal transfer or incomplete lineage sorting) than the majority. We note this is analogous to the distinction people have made for a long time between homology and homoplasy (Mallet, Besansky, et al. 2016; Nixon and Carpenter 2011); in fact, horizontal gene transfer is best viewed as a type of homoplasy.

If there is no majority consensus pattern of lineage descent, then there is no good evidence of a clade, and one is below the SNaRC level given current knowledge. Just like clades at all levels, SNaRCs should not be recognized and named based on a single gene's evidence, but rather on congruence among the majority of gene trees and other types of phylogenetic characters available. SNaRCs are the finest-scale clades named at a given point in time, and they can be counted and used in comparative analyses as long as one keeps in mind the

---

[34]Horizontal genetic transfer can occur in a number of ways, summarized by (Soucy, Huang, et al. 2015). However, it is notable that this usually offers a different phylogeny and recovers different taxa for the gene tree, not the taxon tree (Degnan and Rosenberg 2009).

caveats that: (1) they are not comparable to each other in time depth, or biologically significant properties, or amount of evolutionary change, and (2) they may well be subdivided in the future, given more knowledge. Keeping these warnings in mind, SNaRCs can be used as a better starting point than species for scientific studies of biodiversity, in ways described below.

Using SNaRCs in Systematic, Evolutionary, and Ecological Studies

Studies in such areas as systematics, conservation biology, population genetics, macroevolution, community ecology, and biogeography currently use "species" as a unit of biodiversity in their respective investigations. For example, in ecology, species are used as a stand-in for niche occupiers or trophic nodes. But species are a poor surrogate for biodiversity under an evolutionary worldview. They are at best one level on the tree of life, and there are clades both larger and smaller than named species. Furthermore, a large proportion of currently named species are not units on the tree of life at all, in that they are not monophyletic.[35] All of these factors can potentially confound process studies. Thus phylogenies and rankless phylogenetic taxonomies provide a better theoretical framework than species do for all of these disciplines, and better practical methods for such purposes as conservation assessment as well (Mishler 2010).

Some complain that knowledge in their particular study group is too limited to take a phylogenetic approach to taxonomy. But even if sampling is poor, one can still build a phylogeny with the specimens one has, and this act is no more methodologically suspect than the typical approach to such a situation, which is building a phenogram (a similarity-based classification). Science is about sampling, and one does not need to have sampled every clade in one's group to make the first phylogeny. Phylogenies are hypotheses to be tested by future sampling. Likewise, lacking molecular data is not an excuse to avoid phylogenetic classification including naming SNaRCs. Perfectly good phylogenetic hypotheses can be built using morphological characters. True, it is likely that SNaRCs named with morphological synapomorphies may be particularly subject to subdivision once molecular data are available, but subdivision does not falsify them if they remain monophyletic according to the new data.

Fisher (2006) is a groundbreaking example of in-the-trenches taxonomy using SNaRCs. Her monographic study on a group of tropical mosses relied on all the traditional data-gathering methods: fieldwork making new collections, searching herbaria and literature, measuring morphological characters, sequencing DNA from exemplars, etc. [See Fig. 4.1]. Once she had a phylogeny, she named the best-supported nodes on the tree, as any cautious taxonomist would do. Not all nodes were named; some did not have enough support to be worth naming. For all clades that were named, including the terminal-most clades (i.e., the level formerly treated as species), she provided uninominal PhyloCode-style names. She placed earlier binomial names in synonymy as appropriate, to allow

---

[35]This is to be expected, since monophyly is only a relatively recent desideratum (for some) in taxonomy, many species were named before the cladistics revolution, and a large number of practicing taxonomists still do not accept monophyly as a desideratum.

a connection to the literature and herbaria. For example, the clade *Revolutus* (a SNaRC in our terminology) has *Syrrhopodon revolutus* and *Syrrhopodon microbolax* listed as synonyms. She provided a key to the SNaRCs in the normal manner for practical use in identification. Thus, all the important contributions of a systematic monograph are present for use in other studies.

This example shows it is possible, and indeed more precise, to do standard monographic taxonomy using rankless classification all the way down. As long as we take into account synonymies with traditional taxa, to allow linkages with past literature and databases that use traditional ranked names, there is no barrier to rankless taxonomy. In fact, removing the ranks forces users to consider the nature of clades they compare instead of relying on a false correspondence of ranks.

The taxonomic situation with orangutans provides a nice illustration of the difficulties with ranked classifications. Over the last twenty years there have been debates over the specific status of Sumatran (*Pongo abelii*) and Bornean orangutans (*Pongo pygmaeus*) (Groves 1986; Hobolth, Dutheil, et al. 2011; Xu and Arnason 1996), based primarily on genomic data (Locke, Hillier, et al. 2011). For most of the twentieth century, primatologists identified one orangutan species (*P. pygmaeus*) and several subspecies. Groves rang the change in 1986 when he declared that the *P. p. abelli* and the Sumatran (*P. p. pygmaeus*) "subspecies" were in fact full species, despite reproductive compatibility (adopting a phylogenetic species concept, Groves 2004). This debate continued until the early 2000s (Muir, Galdikas, et al. 2000; Muir, Galdikas, et al. 1998; Muir, Fleming, et al. 2000; Zhang, Ryder, et al. 2001) when a preponderance of molecular, morphological, and ecological data showed that the phylogenetic structure of the two populations was approximately the same in time and degree as between the *Pan troglodytes* (common chimpanzee) and *Pan paniscus* (bonobo) populations (Locke, Hillier, et al. 2011; Muir, Galdikas, et al. 2000; Muir, Galdikas, et al. 1998). However, there are still "subspecies" of *P. pygmaeus* (Caldecott and McConkey 2005):

- the northwest Bornean orangutan, *P. p. pygmaeus;*
- the central Bornean orangutan, *P. p. wurmbii;* and
- the northeast Bornean orangutan, *P. p. mono.*[36]

Moreover, there is considerable haplotype structure within the new species *P. abelii* (Kuhlwilm, de Manuel, et al. 2016). As more data become available, the question of specific rank becomes more urgent under the current paradigm, with concerns about taxonomic inflation. However, a rankless taxonomy, with names applied to SNaRCs supported by a preponderance of gene trees using a concordance-based approach (Steel and Velasco 2014), would resolve much of this concern.

---

[36]A recent paper now assigns specific status, *Pongo tapanuliensis,* to a central Bornean population (Nater, Mattle-Greminger, et al. 2017).

Evolutionary and ecological studies are also better done using clades instead of taxonomic ranks. Diversification studies are often done using species number as a proxy for diversity present in a larger clade (e.g., Alfaro, Santini, et al. 2009), but would be better served by looking at phylogenies down to the SNaRC level, since the number of named species is often not comparable across different groups due to differences in taxonomic customs or amount of taxonomic effort. Rather than comparing species, ecological studies would be more rigorously done comparing clades, as in modern approaches to community phylogenetics (Vamosi, Heard, et al. 2009; Webb, Ackerly, et al. 2002). Clades at any level might be filling a specific niche such as a role in a food web. The important question to be addressed is what level in the tree is filling which ecological role. We should not assume that only named species can do so.

Conservation studies should also be based on phylogenetic approaches that take into account branch lengths on trees rather than species counts per se. Faith (1992) pioneered a concept of phylogenetic diversity (PD) that can be used to characterize biodiversity much more accurately than species number. A related metric has appeared more recently called phylogenetic endemism (PE), which is a measure of how range-restricted lineages are (Rosauer, Laffan, et al. 2009). There is a burgeoning school of research called "spatial phylogenetics" that takes advantage of the availability of large-scale phylogenies, and big distributional data sets derived from museum digitization efforts, to look at patterns of PD and PE on the landscape (Gonzalez-Orozco, Pollock, et al. 2016; Mishler, Knerr, et al. 2014; Nagalingum, Knerr, et al. 2015; Schmidt-Lebuhn, Knerr, et al. 2015; Thornhill, Baldwin, et al. 2017; Thornhill, Mishler, et al. 2016). The results from these studies enable rankless phylogenetic evaluations of conservation priorities, as well as studies of evolution, ecology, and biogeography.

Conclusion

Whether the PhyloCode accepts the SNaRC approach or not, and regardless of how radical it sounds, it is an important step for systematics to take to best account for the theoretical and methodological advances of the last twenty years. As discussed above, it is possible to do all the theoretical and empirical work biological scientists need to do without species or other ranks, and indeed to do it better.

SNaRCs represent the smallest clades for which we have evidence. These are the *infimae* clades. These may or may not coincide with traditional "good species." However, they are the product of good science, and so we should accept them in favor of naïve classifications. While convention may prefer "good species," these are objects of a refined folk taxonomy, not science. For scientists, phenomena such as species are only a starting point for analysis and explanation.

"For the Snark was a Boojum, you see."

–Lewis Carroll, *The Hunting of the Snark* (1876)

ACKNOWLEDGMENTS

BDM thanks Joel Velasco, Kirsten Fisher, Nico Cellinese, and David Baum for many discussions on these issues, members of the audience of the "Species in the Age of Discordance" Conference (23–25 March 2017) at the University of Utah, and acknowledges support of NSF grant DEB-1354552. JSW thanks members of the audience of the Combined Invertebrate Biodiversity & Conservation/Society of Australian Systematic Biologists Conference (4–7 December 2011), and the "Species Delimitation in the Age of Genomics" conference, Centre for Biodiversity Analysis, ANU (28–30 April 2015) for discussion; John Hawks for advice on orangutans; and Malte Ebach for many discussions on this subject.

LITERATURE CITED

Agapow, Paul–Michael, Olaf R. P. Bininda–Emonds, Keith A. Crandall, John L. Gittleman, Georgina M. Mace, Jonathon C. Marshall, and Andy Purvis. 2004. "The Impact of Species Concept on Biodiversity Studies." *The Quarterly Review of Biology* 79 (2): 161–179.

Agassiz, Louis. 1859. *An Essay on Classification*. London: Longman, Brown, Green, Longmans and Roberts and Trübner.

Alfaro, Michael E., Francesco Santini, Chad Brock, Hugo Alamillo, Alex Dornburg, Daniel L. Rabosky, Giorgio Carnevale, and Luke J. Harmon. 2009. "Nine Exceptional Radiations Plus High Turnover Explain Species Diversity in Jawed Vertebrates." *Proceedings of the National Academy of Sciences* 106 (32): 13410–13414.

Amitani, Yuichi. 2015. "Prototypical Reasoning About Species and the Species Problem." *Biological Theory* 10 (4): 289–300.

Anon. 1908. "Discussion of the Species Question." *The American Naturalist* 42 (496): 272–281.

Apel, Jochen, Monika Dullstein, and Pawel Radchenko. 2009. "Data-Phenomena-Theories: What's the Notion of a Scientific Phenomenon Good For?" *Journal for General Philosophy of Science* 40 (1): 125–128.

Atran, Scott. 1990. *The Cognitive Foundations of Natural History*. New York: Cambridge University Press.

Atran, Scott. 1999. "The Universal Primacy of Generic Species in Folkbiological Taxonomy: Implications for Human Biological, Cultural and Scientific Evolution." In *Species, New Interdisciplinary Essays*, edited by R. A. Wilson. Cambridge, MA: Bradford/MIT Press.

Baum, David A. 2009. "Species as Ranked Taxa." *Systematic Biology* 58 (1): 74–86.

Beltrán, Margarita, Chris D. Jiggins, Vanessa Bull, Mauricio Linares, James Mallet, W. Owen McMillan, and Eldredge Bermingham. 2002. "Phylogenetic Discordance at the Species Boundary: Comparative Gene Genealogies Among Rapidly Radiating *Heliconius* Butterflies." *Molecular Biology and Evolution* 19 (12): 2176–2190.

Berlin, Brent. 1973. "Folk Systematics in Relation to Biological Classification and Nomenclature." *Annual Review of Ecology and Systematics* 4 (1): 259–271.

Berlin, Brent. 1976. "The Concept of Rank in Ethnobiological Classification: Some Evidence From Aguaruna Folk Botany." *American Ethnologist* 3 (3): 381–399.

Berlin, Brent, Dennis E. Breedlove, and Peter H. Raven. 1973. "General Principles of Classification and Nomenclature in Folk Biology." *American Anthropologist* 75 (1): 214–242.

Birky, Jr., C. William, Joshua Adams, Marlea Gemmel, and Julia Perry. 2010. "Using Population Genetic Theory and DNA Sequences for Species Detection and Identification in Asexual Organisms." *PLoS ONE* 5 (5): e10609.

Blaxter, Mark, Jenna Mann, Tom Chapman, Fran Thomas, Claire Whitton, Robin Floyd, and Eyualem Abebe. 2005. "Defining Operational Taxonomic Units Using DNA Barcode Data." *Philosophical Transactions: Biological Sciences* 360 (1462): 1935–1943.

Bogart, James P. 2003. "Genetics and Systematics of Hybrid Species." *Reproductive Biology and Phylogeny of Urodela* 1:109–134.

Bogen, James, and James Woodward. 1988. "Saving the Phenomena." *The Philosophical Review* 67 (3): 303–352.

Breidbach, Olaf, and Michael T. Ghiselin. 2006. "Athanasius Kircher (1602–1680) on Noah's Ark: Baroque "Intelligent Design" Theory." *Proceedings of the California Academy of Science* 57 (36): 991–1002.

Buteo, Johannes. 1554. *Opera Geometrica: De Arca Noe, De Sublicio Ponte Caesaris, Confutatio Quadraturae Circuli Ab Orontio Finaeo Factae Etc*. Lugduni: Thomas Bertellus.

Caldecott, Julian, and Kim McConkey. 2005. "Orangutan Overview." Chap. 9 in *World Atlas of Great Apes and Their Conservation*. Berkeley and Los Angeles, CA: University of California Press.

Cantino, Philip D., and Kevin de Queiroz. 2000. *PhyloCode: International Code of Phylogenetic Nomenclature*. Web Page. https://www.ohio.edu/phylocode/.

Cellinese, Nico, David A. Baum, and Brent D. Mishler. 2012. "Species and Phylogenetic Nomenclature." *Systematic Biology* 61 (5): 885–891.

Claridge, Michael F., H. A. Dawah, and M. R. Wilson. 1997. *Species: The Units of Biodiversity*. xvi, 439. London; New York: Chapman & Hall.

De Queiroz, Kevin. 1999. "The General Lineage Concept of Species and the Defining Properties of the Species Category." In *Species, New Interdisciplinary Essays*, edited by R. A. Wilson. Cambridge, MA: Bradford/MIT Press.

De Queiroz, Kevin. 2005. "Ernst Mayr and the Modern Concept of Species." *Proceedings of the National Academy of Sciences* 102 (Supp. 1): 6600–6607.

De Queiroz, Kevin. 2007. "Species Concepts and Species Delimitation." *Systematic Biology* 56 (6): 879–886.

De Queiroz, Kevin, and Jacques Gauthier. 1992. "Phylogenetic Taxonomy." *Annual Review of Ecology and Systematics* 23 (1): 449–480.

De Queiroz, Kevin, and Jacques Gauthier. 1994. "Toward a Phylogenetic System of Biological Nomenclature." *Trends in Ecology & Evolution* 9 (1): 27–31.

Decock, Lieven, and Igor Douven. 2011. "Similarity After Goodman." *Review of Philosophy and Psychology* 2 (1): 61–75.

Degnan, J., and N. Rosenberg. 2009. "Gene Tree Discordance, Phylogenetic Inference and the Multispecies Coalescent." *Trends in Ecology and Evolution* 24 (6): 332–340.

Després, Laurence, Emeline Pettex, Valérie Plaisance, and François Pompanon. 2002. "Speciation in the Globeflower Fly *Chiastocheta* Spp. (Diptera: Anthomyiidae) in Relation to Host Plant Species, Biogeography, and Morphology." *Molecular Phylogenetics and Evolution* 22 (2): 258–268.

Dobzhansky, Theodosius. 1935. "A Critique of the Species Concept in Biology." *Philosophy of Science* 2 (3): 344–355.

Dobzhansky, Theodosius. 1937. "What Is a Species?" *Scientia* 61:280–286.

Dupré, John. 1981. "Natural Kinds and Biological Taxa." *The Philosophical Review* 90 (1): 66–90.

Durkheim, Emile, and Marcel Mauss. 1963. *Primitive Classification*. 96. London: Cohen & West.

Ereshefsky, Marc. 1999. "Species and the Linnaean Hierarchy." In *Species, New Interdisciplinary Essays*, edited by R. A. Wilson. Cambridge, MA: Bradford/MIT Press.

Faith, Daniel P. 1992. "Conservation Evaluation and Phylogenetic Diversity." *Biological Conservation* 61 (1): 1–10.

Fisher, Kirsten M. 2006. "Rank-Free Monography: A Practical Example From the Moss Clade Leucophanella (Calymperaceae)." *Systematic Botany* 31 (1): 13–30.

Fitzhugh, Kirk. 2005. "The Inferential Basis of Species Hypotheses: The Solution to Defining the Term 'Species'." *Marine Ecology* 26 (3–4): 155–165.

Fitzhugh, Kirk. 2009. "Species as Explanatory Hypotheses: Refinements and Implications." *Acta Biotheoretica* 57 (1): 201–248.

Gayon, Jean. 1996. "The Individuality of the Species: A Darwinian Theory? – From Buffon to Ghiselin, and Back to Darwin." *Biology & Philosophy* 11:215–244.

Ghiselin, Michael T. 1974. "A Radical Solution to the Species Problem." *Systematic Zoology* 23:536–544.

Ghiselin, Michael T. 1997. *Metaphysics and the Origin of Species*. Albany: State University of New York Press.

Gonzalez-Orozco, Carlos E., Laura J. Pollock, Andrew H. Thornhill, Brent D. Mishler, Nunzio Knerr, Shawn W. Laffan, Joseph T. Miller, et al. 2016. "Phylogenetic Approaches Reveal Biodiversity Threats Under Climate Change." *Nature Climate Change* 6 (December): 1110–1114.

Goodman, Nelson. 1972. *Problems and Projects*. Indianapolis: Bobbs-Merrill.

Green, David M. 2005. "Designatable Units for Status Assessment of Endangered Species." *Conservation Biology* 19 (6): 1813–1820.

Groves, C. 2004. "The What, Why and How of Primate Taxonomy." *International Journal of Primatology* 25 (5): 1105–1126.

Groves, Colin P. 1986. "Systematics of the Great Apes." In *Comparative Primate Biology*, edited by D. R. Swindler and J. Erwin, vol. 1: Systematics, Evolution, and Anatomy. New York: Alan R. Liss.

Hacking, Ian. 1983. *Representing and Intervening: Introductory Topics in the Philosophy of Natural Science*. Cambridge UK: Cambridge University Press.

Hacking, Ian. 1990. "Natural Kinds." In *Perspectives on Quine*, edited by R. B. Barrett and R. F. Gibson. Cambridge: Blackwell.

Hacking, Ian. 2007. "Natural Kinds: Rosy Dawn, Scholastic Twilight." *Royal Institute of Philosophy Supplement* 82 (Supplement 61): 203–239.

Hausdorf, Bernhard. 2011. "Progress Toward a General Species Concept." *Evolution and Development* 65 (4): 923–931.

Hey, Jody. 2001. "The Mind of the Species Problem." *Trends in Ecology and Evolution* 16 (7): 326–329.

Hey, Jody. 2006. "On the Failure of Modern Species Concepts." *Trends in Ecology and Evolution* 21 (8): 447–450.

Hobolth, Asger, Julien Y. Dutheil, John Hawks, Mikkel H. Schierup, and Thomas Mailund. 2011. "Incomplete Lineage Sorting Patterns Among Human, Chimpanzee, and Orangutan Suggest Recent Orangutan Speciation and Widespread Selection." *Genome Research* 21 (3): 349–356.

Hull, David L. 1975. "The Ontological Status of Species as Evolutionary Units." In *Foundational Problems in the Special Sciences: Proceedings of the Fifth International Congress of Logic, Methodology, and Philosophy of Science, London, Ontario, Canada*, edited by R. E. Butts and J. Hintikka. Dordrecht, Holland; Boston: D. Reidel.

Hull, David L. 1976. "Are Species Really Individuals?" *Systematic Zoology* 25:174–191.

Isaac, N. J. B., J. Mallet, and G. M. Mace. 2004. "Taxonomic Inflation: Its Influence on Macroecology and Conservation." *Trends in Ecology and Evolution* 19 (9): 464–469.

Khalidi, Muhammad Ali. 2013. *Natural Categories and Human Kinds: Classification in the Natural and Social Sciences*. Cambridge University Press.

Kircher, Athanasius. 1675. *Arca Noë in Tres Libros Digesta*. Amsterdam: Joannem Janssonium à Waesberge.

Kuhlwilm, Martin, Marc De Manuel, Alexander Nater, Maja P. Greminger, Michael Krützen, and Tomas Marques-Bonet. 2016. "Evolution and Demography of the Great Apes." *Current Opinion in Genetics and Development* 41:124–129.

Levin, Donald A. 1979. "The Nature of Plant Species." *Science* 204 (4391): 381–384.

Locke, Devin P., Ladeana W. Hillier, Wesley C. Warren, Kim C. Worley, Lynne V. Nazareth, Donna M. Muzny, Shiaw-Pyng Yang, et al. 2011. "Comparative and Demographic Analysis of Orang-Utan Genomes." *Nature* 469 (7331): 529–533.

Lodé, Thierry. 2013. "Adaptive Significance and Long-Term Survival of Asexual Lineages." *Evolutionary Biology* 40 (3): 450–460.

Mallet, James, Nora Besansky, and Matthew W. Hahn. 2016. "How Reticulated Are Species?" *BioEssays* 38 (2): 140–149.

Massimi, Michela. 2008. "Why There Are No Ready-Made Phenomena: What Philosophers of Science Should Learn From Kant." *Kant and Philosophy of Science Today, Royal Institute of Philosophy Supplement* 63:1–35.

Massimi, Michela. 2011. "From Data to Phenomena: A Kantian Stance." *Synthese* 182 (1): 101–116.

Mayr, Ernst. 1996. "What Is a Species, and What Is Not?" *Philosophy of Science* 2:262–277.

McOuat, Gordon R. 2001. "Cataloguing Power: Delineating 'Competent Naturalists' and the Meaning of Species in the British Museum." *The British Journal for the History of Science* 34:1–28.

Medin, Douglas L., and Scott Atran. 1999. *Folkbiology*. Cambridge MA: MIT Press.

Mishler, Brent D. 2009. "Three Centuries of Paradigm Changes in Biological Classification: Is the End in Sight?" *Taxon* 58 (1): 61–67.

Mishler, Brent D. 1999. "Getting Rid of Species?" In *Species, New Interdisciplinary Essays*, edited by R. A. Wilson. Cambridge, MA: Bradford/MIT Press.

Mishler, Brent D. 2010. "Species Are Not Uniquely Real Biological Entities." Chap. 6 in *Contemporary Debates in Philosophy of Biology*, edited by F. J. Ayala and R. Arp. Chichester: Wiley-Blackwell.

Mishler, Brent D., and Robert N. Brandon. 1987. "Individuality, Pluralism, and the Phylogenetic Species Concept." *Biology & Philosophy* 2:397–414.

Mishler, Brent D., and Michael J. Donoghue. 1982. "Species Concepts: A Case for Pluralism." *Systematic Zoology* 31:491–503.

Mishler, Brent D., Nunzio Knerr, Carlos E. González-Orozco, Andrew H. Thornhill, Shawn W. Laffan, and Joseph T. Miller. 2014. "Phylogenetic Measures of Biodiversity and Neo- And Paleo-Endemism in Australian *Acacia*." *Nature Communications* 5.

Mishler, Brent D., and Edward C. Theriot. 2000. "The Phylogenetic Species Concept (Sensu Mishler and Theriot): Monophyly, Apomorphy, and Phylogenetic Species Concepts." In *Species Concepts and Phylogenetic Theory: A Debate*, edited by Q. D. Wheeler and R. Meier. New York: Columbia University Press.

Moritz, Craig, and Ke Bi. 2011. "Spontaneous Speciation by Ploidy Elevation: Laboratory Synthesis of a New Clonal Vertebrate." *Proceedings of the National Academy of Sciences* 108 (24): 9733–9734.

Muir, C. Cam, B. M. F. Galdikas, and Andrew T. Beckenbach. 2000. "mtDNA Sequence Diversity of Orangutans From the Islands of Borneo and Sumatra." *Journal of Molecular Evolution* 51 (5): 471–480.

Muir, C. Cam, Birute M. F. Galdikas, and Andrew T. Beckenbach. 1998. "Is There Sufficient Evidence to Elevate the Orangutan of Borneo and Sumatra to Separate Species?" *Journal of Molecular Evolution* 46 (4): 378–379.

Muir, Graham, Colin Fleming, and Christian Schlötterer. 2000. "Species Status of Hybridizing Oaks." *Nature* 405:1016.

Nagalingum, Nathalie S., Nunzio Knerr, Shawn W. Laffan, Carlos E. González-Orozco, Andrew H. Thornhill, Joseph T. Miller, and Brent D. Mishler. 2015. "Continental Scale Patterns and Predictors of Fern Richness and Phylogenetic Diversity." *Frontiers in Genetics* 6 (132).

Naomi, Shun-Ichiro. 2011. "On the Integrated Frameworks of Species Concepts: Mayden's Hierarchy of Species Concepts and De Queiroz's Unified Concept of Species." *Journal of Zoological Systematics and Evolutionary Research* 49 (3): 177–184.

Nater, Alexander, Maja P. Mattle-Greminger, Anton Nurcahyo, Matthew G. Nowak, Marc De Manuel, Tariq Desai, Colin Groves, et al. 2017. "Morphometric, Behavioral, and Genomic Evidence for a New Orangutan Species." *Current Biology* 27 (22): 3487–3498.e10.

Nixon, Kevin C., and James M. Carpenter. 2011. "On Homology." *Cladistics* 28 (2): 160–169.

Ochman, Howard, Emmanuelle Lerat, and Vincent Daubin. 2005. "Examining Bacterial Species Under the Specter of Gene Transfer and Exchange." *Proceedings of the National Academy of Sciences* 102 (Suppl 1): 6595–6599.

Oreskes, Naomi, and Homer E. Legrand. 2003. *Plate Tectonics: An Insider's History of the Modern Theory of the Earth*. Boulder, Colo.: Westview Press.

Pleijel, Frederik. 1999. "Phylogenetic Taxonomy, a Farewell to Species, and a Revision of *Heteropodarke* (*Hesionidae, Polychaeta, Annelida*)." *Systematic Biology* 48 (4): 755–789.

Pleijel, Frederik, and G. W. Rouse. 2000. "Least-Inclusive Taxonomic Unit: A New Taxonomic Concept for Biology." *Proceedings of the Royal Society of London - Series B: Biological Sciences* 267 (1443): 627–630.

Reydon, Thomas A. C. 2005. *Species as Units of Generalization in Biological Science: A Philosophical Analysis*. 153. Rotterdam: Self-published.

Rieppel, Olivier. 2010. "New Essentialism in Biology." *Philosophy of Science* 77 (5): 662–673.

Rieseberg, Loren H. 1997. "Hybrid Origins of Plant Species." *Annual Review of Ecology and Systematics* 28:359–389.

Rosauer, Dan F., Shawn W. Laffan, Michael D. Crisp, Stephen C. Donnellan, and Lyn G. Cook. 2009. "Phylogenetic Endemism: A New Approach for Identifying Geographical Concentrations of Evolutionary History." *Molecular Ecology* 18 (19): 4061–4072.

Scerri, Eric R. 2007. *The Periodic Table: Its Story and Its Significance*. New York: Oxford University Press.

Schindler, Samuel. 2011. "Bogen and Woodward's Data-Phenomena Distinction, Forms of Theory-Ladenness, and the Reliability of Data." *Synthese* 182 (1): 39–55.

Schmidt-Lebuhn, Alexander N., Nunzio J. Knerr, Joseph T. Miller, and Brent D. Mishler. 2015. "Phylogenetic Diversity and Endemism of Australian Daisies (Asteraceae)." *Journal of Biogeography* 42 (6): 1114–1122.

Sokal, Robert R., and T. Crovello. 1970. "The Biological Species Concept: A Critical Evaluation." *American Naturalist* 104:127–153.

Soucy, Shannon M., Jinling Huang, and Johann Peter Gogarten. 2015. "Horizontal Gene Transfer: Building the Web of Life." *Nature Reviews Genetics* 16 (8): 472–482.

Sousa, Paulo, Scott Atran, and Douglas Medin. 2002. "Essentialism and Folkbiology: Evidence From Brazil." *Journal of Cognition and Culture* 2 (3): 195.

Staley, James T. 2006. "The Bacterial Species Dilemma and the Genomic – Phylogenetic Species Concept." *Philosophical Transactions of the Royal Society B: Biological Sciences* 361 (1475): 1899–1909.

Staley, James T. 2013. "Transitioning Toward a Universal Species Concept for the Classification of All Organisms." In *The Species Problem - Ongoing Issues*, edited by I. Ya. Pavlinov.

Steel, Mike, and Joel D. Velasco. 2014. "Axiomatic Opportunities and Obstacles for Inferring a Species Tree From Gene Trees." *Systematic Biology* 63 (5): 772–778.

Thornhill, Andrew H., Bruce G. Baldwin, William A. Freyman, Sonia Nosratinia, Matthew M. Kling, Naia Morueta-Holme, Thomas P. Madsen, David D. Ackerly, and Brent D. Mishler. 2017. "Spatial Phylogenetics of the Native California Flora." *BMC Biology* 15 (1): 96.

Thornhill, Andrew H., Brent D. Mishler, Nunzio J. Knerr, Carlos E. González-Orozco, Craig M. Costion, Darren M. Crayn, Shawn W. Laffan, and Joseph T. Miller. 2016. "Continental-Scale Spatial Phylogenetics of Australian Angiosperms Provides Insights Into Ecology, Evolution and Conservation." *Journal of Biogeography* 43 (11): 2085–2098.

Turner, George F. 2002. "Parallel Speciation, Despeciation and Respeciation: Implications for Species Definition." *Fish and Fisheries* 3 (3): 225–229.

Vamosi, Steven M., Stephen B. Heard, Jana C. Vamosi, and Campbell O. Webb. 2009. "Emerging Patterns in the Comparative Analysis of Phylogenetic Community Structure." *Molecular Ecology* 18 (4): 572–592.

Van Valen, L. 1976. "Ecological Species, Multispecies, and Oaks." *Taxon* 25:233–239.

Vanderlaan, Tegan A., Malte C. Ebach, David M. Williams, and John S. Wilkins. 2013. "Defining and Redefining Monophyly: Haeckel, Hennig, Ashlock, Nelson and the Proliferation of Definitions." *Australian Systematic Botany* 26 (5): 347–355.

Vrana, P., and Ward Wheeler. 1992. "Individual Organisms as Terminal Entities: Laying the Species Problem to Rest." *Cladistics* 8:67–72.

Webb, Campbell O., David D. Ackerly, Mark A. McPeek, and Michael J. Donoghue. 2002. "Phylogenies and Community Ecology." *Annual Review of Ecology and Systematics* 33:475–505.

Wheeler, Quentin D., and Norman I. Platnick. 2000. "The Phylogenetic Species Concept (Sensu Wheeler and Platnick)." In *Species Concepts and Phylogenetic Theory: A Debate*, edited by Quentin D. Wheeler and Rudolf Meier. New York: Columbia University Press.

Wilkins, John. 2013. "We Have 'Species' Thanks to Noah's Ark." *The Conversation* (28 October 2013).

Wilkins, John S. 2007a. "The Concept and Causes of Microbial Species." *Studies in History and Philosophy of the Life Sciences* 28 (3): 389–408.

Wilkins, John S. 2007b. "The Dimensions, Modes and Definitions of Species and Speciation." *Biology & Philosophy* 22 (2): 247–266.

Wilkins, John S. 2009. *Species: A History of the Idea*. Berkeley: University of California Press.

Wilkins, John S. 2010. "What Is a Species? Essences and Generation." *Theory in Biosciences* 129:141–148.

Wilkins, John S. 2011. "Philosophically Speaking, How Many Species Concepts Are There?" *Zootaxa* 2765:58–60.

Wilkins, John S. 2013a. "Biological Essentialism and the Tidal Change of Natural Kinds." *Science and Education* 22, no. 2 (February): 221–240.

Wilkins, John S. 2013b. "Essentialism in Biology." In *Philosophy of Biology: A Companion for Educators*, edited by Kostas Kampourakis. Dordrecht: Springer.

Wilkins, John S. 2018. *Species: The Evolution of the Idea*. 2nd ed. Boca Raton, FL: CRC Press.

Wilkins, John S., and Malte C. Ebach. 2013. *The Nature of Classification: Kinds and Relationships in the Natural Sciences.* London: Palgrave Macmillan.

Woodward, James. 2000. "Data, Phenomena, and Reliability." *Philosophy of Science* 67 (3): 163–179.

Xu, Xiufeng, and Ulfur Arnason. 1996. "The Mitochondrial DNA Molecule of Sumatran Orangutan and a Molecular Proposal for Two (Bornean and Sumatran) Species of Orangutan." *Journal of Molecular Evolution* 43 (5): 431–437.

Zachar, Peter. 2000. "Folk Taxonomies Should Not Have Essences, Either: A Response to the Commentary." *Philosophy, Psychiatry, and Psychology* 7 (3): 191–194.

Zachos, Frank E., and Sandro Lovari. 2013. "Taxonomic Inflation and the Poverty of the Phylogenetic Species Concept – A Reply to Gippoliti and Groves." *Hystrix: The Italian Journal of Mammalogy* 24 (2): 142–144.

Zhang, Yun-Wu, Oliver A. Ryder, and Ya-Ping Zhang. 2001. "Genetic Divergence of Orangutan Subspecies (*Pongo pygmaeus*)." *Journal of Molecular Evolution* 52 (6): 516–526.

5 Discussion
What Would the World Be Like without the Species Rank?

Even if it is accepted that the species level is arbitrary, as argued initially and most cogently by Darwin (see Mishler 2010 for an explication of Darwin's argument), the case is not closed when one gives up the species rank. Given how importantly the species rank has featured in academic biology as well as in conservation and management, one cannot just reject species without providing a replacement view. People need to get on with their work. So, how to proceed in ecology and evolutionary biology? What interacts in an ecosystem? What evolves? What are the units of biodiversity? How can people interested in more practical issues, like conservation, biodiversity inventories, field guides, monographs, restoration ecology, etc. move forward?

IMPLICATIONS FOR STUDIES OF ECOLOGY

Species have figured prominently in ecological theory and practice. They have widely been considered the actors in the "Ecological Theater and the Evolutionary Play" (Hutchinson 1965). Ecologists have traditionally treated infra-specific interactions to be different in kind than inter-specific interactions. So... if we are agreed that the species rank is meaningless, then can we still do ecology? Of course! In fact, we can develop a richer and more accurate view of ecology by considering clades at different levels as potential ecological interactors. There was never any reason to treat the species level as some kind of a magical dividing line. The view advanced in this book frees up ecology to be done in a more realistic multi-level manner. In truth there are clades interacting in various ways at a variety of nested scales, none of which are "primary." The level at which ecological interactions change should be left as an open question, as interesting hypotheses to test, rather than making an a priori assumption that the species boundary is always the answer.

Therefore, ecological interchangeability of organisms is seen to be an empirical question, not a theoretical one. Often very fine-scale populations, much smaller than anyone would want to call "species" are found to be locally adapted physiologically and are definitely not interchangeable with other members of the same-named species. This is the basis of recent arguments in restoration ecology to use local genotypes (McKay et al. 2005, Gustafson et al. 2014), and there is a whole industry developed to supply this demand. Other times large clades such as cacti have conservative habitat preferences that allow them to be assigned to a distinctive niche at a high level. The level in a phylogeny where ecological traits change, and why, is a tractable trait reconstruction problem – sorting out evolutionarily rapid from evolutionarily conservative ecology is one of the most fascinating areas of study in biology today.

The implications for ecology go on and on, but the answers are similar in each case. For example, studying climate change responses at the species level, while by far the most common approach, can now be seen to be quite limited. Species are unlikely to respond to climate change as a whole entity. Their physiology is often genetically variable across their range. Which of course is a good thing: local populations will (hopefully) adapt at the margins of their clade's range. So-called "assisted migration," controversial for several reasons, suffers from this problem as well: care needs to be taken when deciding which genotypes of a named species should be moved, if any.

Likewise, food webs are not best thought of as exclusively comprised of species-level entities. Sometimes the relevant players in food webs are local populations, like a particular salmon run; other times they are much deeper clades, like diatoms. Ecological action is a contingent question and needs to be approached with an open mind. There certainly are ecological actors in Hutchinson's sense; however, they are most often not the named species but are instead clades at finer or deeper levels in the tree of life.

IMPLICATIONS FOR STUDIES OF "SPECIATION" (="DIVERSIFICATION")

If we are agreed that the species rank is meaningless, then another controversial implication is that there is no such process as "speciation." Many would object, of course, since they have labeled their entire research program as studies of "speciation!" But their field does not disappear, it actually becomes richer. Just like ecology, you can still study evolutionary divergence without species; in fact you can do it better without wearing species blinders. There are lineages diverging from each other (and sometimes coming back together) at *many* levels in evolution. No one level is "most important" or privileged in any way. Different processes are involved at different levels, and to fully understand what is going on in the diversification of a group, you need to look at all the levels, not just concentrate on the one labeled by traditional taxonomists as the species level. This remains an important field, but *diversification* is the right name to use for it, rather than *speciation*. There needs to be both an appreciation for multiple levels of divergence, and an appreciation that gene flow (or lack thereof) is not the only process to consider.

The process of studying diversification should always start with a phylogeny down to as fine a level as is possible to resolve with the data at hand (i.e., the SNaRCs). To understand the "cutting edge" of diversification, one should compare the terminal-most sister clades to each other to see which traits (i.e., genomic, morphological, physiological, geographic, developmental, reproductive, etc.) are reconstructed to change on the branches separating them. Those characters are the candidates for possible causes or constraints affecting diversification in that particular sister pair. If one looks at many terminal sister pairs across the same phylogeny in this way, then one can look for common denominators to see whether there is some dominant process operating in that particular group.

For example, if the terminal sister pairs differ most often by the acquisition of some reproductive isolating mechanism in one or both of the pair, that is prima facie evidence that cessation of interbreeding is indeed the most important factor driving primary divergence in that particular group. Or, if terminal sister pairs differ most often by the acquisition of a novel ecological niche in one of the lineages, that is prima facie evidence that ecology is the most important factor in primary divergence in that particular group. Or, if terminal sister pairs differ most often by geographic isolation that is prima facie evidence that allopatry is the most important factor in primary divergence in that particular group. And so on for different combinations of processes.

Even though the terminal-most sister pairs provide the best evidence for inferring the "cutting edge" of recent divergence, one can and should, of course, also compare sister lineages deeper in the tree to make inferences about processes affecting earlier stages of evolutionary divergence. The age-old debate, stemming back to the different viewpoints of Darwin and Wallace (discussed in Grant 1981), as to whether reproductive isolation arises early and is the cause of later ecological difference (Wallace) or vice-versa (Darwin) can be settled in this way, at least for particular groups. Other debates such as the role of different modes of geographic isolation (Mayr 1982) can be settled in this manner also.

Thus it is not necessary to decide *a priori* which of the many proposed species concepts might best capture mechanisms featuring in divergence of one's study group. We can and should initially be neutral about mechanisms to begin with, and let the data reveal what processes are most likely operating. The SNaRC approach allows us to do this: build phylogenies first, then map inferences of changes in mechanisms second. One should name taxa last, taking into account what has been discovered in this neutral manner. This stands the current field of "speciation" on its head, since the usual procedure now is to begin with traditional species classifications and then try to figure out how those entities differ (e.g., Doyle et al. 2004)!

IMPLICATIONS FOR STUDIES OF EVOLUTION

The view of species advocated here also would imply that neither the common distinction between micro- and macroevolution nor the common distinction between population genetics and systematics (e.g., Avise 1989) – using the species level as the boundary – make sense. The species level should not be viewed as a magical dividing line here either. As discussed earlier, the processes of divergence and reticulation don't respect species boundaries; both are, in fact, happening within and among named species. The preponderance of one versus the other may change at different scales, but this is another contingent issue worthy of study rather than the a priori assumption that divergence only happens above the species level and reticulation only happens below (Nixon & Wheeler 1990).

Diversification is often used in a different sense these days than advocated in the previous section. It is often taken to mean the net accumulation of species in a clade, i.e., speciation minus extinction (e.g., Nürk et al. 2020). However, taking the hierarchical view of lineages I am arguing for here, it would be better to use *diversification* to refer to splitting of lineages at various levels, *extinction* to refer to the loss of lineages at various levels, and *diversity* (or richness) to refer to the net number of lineages at some level present at a given time. So, *diversity = diversification – extinction*. Diversification and extinction are process terms, while diversity in this sense is just a metric summarizing the current situation. There are other, perhaps more useful, phylogenetic metrics of diversity as well; see below.

One of the worst metrics of diversity to use is the number of named species (or genera) in a clade, but unfortunately, there is a tendency in the literature to use just this measure (e.g., Nürk et al. 2020). There are a number of problems with this, most important of which are the arguments developed above for why named species taxa are not, and cannot be, comparable entities even with ideal knowledge. Furthermore, there are artifacts incorporated in this metric given the less than ideal situation in most cases: different major groups have had very different amounts of taxonomic study and the number of named species is likely just to be a reflection of how well-studied a group is. There are also quite different traditions of splitting and lumping in communities of taxonomists who work on different major groups. Many traditional taxonomists do not take monophyly as an important criterion, so many named species were not even *intended* to be minimal clades. So given a simple observation of 20 named species listed for one clade, and 200 listed for its sister clade, you don't really know how biodiversity compares between the two.

The neutral approach discussed in the previous section is useful here as well – to compare the diversity in different clades objectively one should build a phylogeny down to SNaRCs. Comparing the number of SNaRCs among sister groups gives you a rough estimate of the net amount of lineage-splitting that is represented in each. However, given the potential for artifacts discussed above, it is important to ensure that about the same amount of sampling and the same level of care have gone into reconstructing the phylogeny of the sister groups being compared.

The number of SNaRCs (i.e., *richness*) can thus tell you something about evolutionary patterns, but there is much more information to be gained from the phylogeny than simply counting the number of known tips. The branches connecting the tips are important also, and adding inferences about them adds depth to the evolutionary comparisons that can be made. The distribution of branch lengths on the tree tells you about the "shape" of evolution (Gould et al. 1987). Furthermore, there are multiple ways to represent branch lengths on a tree; thus evolution has more than one shape at the same time, depending on how you measure it. Faith (1992) pioneered an evolutionary metric of biodiversity called phylogenetic diversity (PD), which sums up the branch lengths connecting the tips of a tree that are present in a clade, or in a geographic location, or co-occupying an ecological community.

There are different dimensions to phylodiversity. As discussed by Kling et al. (2018), given a fixed phylogenetic topology, there are three main types of representations of branch lengths, yielding three distinct facets of phylodiversity that can be used to assess biodiversity in a location using PD: (1) if branches are scaled to character change (resulting in a *phylogram*), then PD is a measure of trait disparity; (2) if branches are scaled to time (resulting in a *chronogram*), then PD is a measure of evolutionary time elapsed; and (3) if all branches are scaled to equal length (resulting in a *cladogram*), then PD is a measure of the net amount of lineage-splitting. Each of these facets, and more besides, can be measured using PD, which complements the simple metric of richness and adds multiple dimensions to studies of the origin and maintenance of biodiversity. For example, comparing PD between sister clades is much more informative for studies of diversification than simply comparing the number of named taxa they contain (Miller et al. 2018). Comparing ecological communities using measures of PD is much more informative than simply comparing how many species each contains (Webb et al. 2002). Yet another place where the power of this approach is demonstrated is when you place different measures of PD on a map, which has led to the development of a new field of biogeography as outlined in the following section.

IMPLICATIONS FOR THE STUDY OF BIOGEOGRAPHY

To summarize the previous section, our perception of the patterns of biodiversity, and their ecological and evolutionary significance, is vastly enhanced when phylogenetic branch lengths are added to simple measure of richness (whether of SNaRCs or of traditional species). Alpha-diversity has usually been measured geographically by examining changes in the number of species across a region to identify areas of particularly high species richness and endemism. Beta-diversity, or turn-over on the landscape, is likewise typically measured by comparing proportions of species shared among subareas of a region. Adding a phylogenetic approach by considering branch lengths that connect SNaRCs greatly improves studies of both alpha- and beta-diversity.

Recent developments in this area have led to the development of a new field that we call *spatial phylogenetics*. This is "big data" research, integrating massive distributional datasets (largely derived from recent digitizing efforts in museum and herbarium collections), rapidly accumulating molecular datasets (e.g., cheaper DNA sequencing feeding the exponential growth of Genbank), and new computational methods allowing reconstruction of very large phylogenies. Using phylogenies encompassing whole biotas allows you to put evolutionary history on a map in a GIS context, opening up many avenues of spatially explicit studies.

Building on the concept of *phylogenetic diversity* (PD, i.e., the portion of a global phylogeny found in a local area), Rosauer et al. (2009) developed a new metric called *phylogenetic endemism* (PE), that weights the contribution to PD of each branch inversely by how widespread it is. Mishler et al. (2014) developed additional phylogenetic tools, including two new metrics, *relative phylogenetic diversity* (RPD) and *relative phylogenetic endemism* (RPE), and a new spatial randomization statistical test called *categorical analysis of neo-and paleo-endemism* (CANAPE), that classifies different types of endemism. RPD and RPE are ratios that compare PD or PE (respectively) measured on the original tree (either a phylogram or a chronogram) with PD or PE (respectively) measured on a comparison tree where all branches are of equal length (i.e., a cladogram); thus they allow the discovery of significant geographic concentrations of long or short branches. Laffan et al. (2016) took the next step and looked at beta-diversity in PD and PE, developing a really useful metric of phyloturnover that emphasizes lineages that are rare on the map called *phylogenetic range-weighted turnover* (PhyloRWT).

These spatial phylogenetic methods are all rankless, since it does not matter what taxonomic level the terminals of the phylogeny represent, as long as they are monophyletic and their geographic distribution can be characterized, and are thus relatively robust to lumping and splitting decisions by taxonomists. Hotspots of diversity and endemism can be mapped, their make-up assessed, and similarities and differences among them characterized. Using hypothesis tests based on spatial randomization, insights can be gained into ecological, evolutionary, and biogeographic processes that have shaped these patterns. Furthermore, understanding such patterns of biodiversity on the landscape is also important for conservation planning, given the need to prioritize efforts in the face of rapid habitat loss and human-induced climate change (see next section). Note that all these insights into biodiversity are gained by directly using the properties of phylogenies themselves, rather than counting species or other taxa. Thus spatial phylogenetics provides one of the main empirical ways forward in a world without the species rank.

IMPLICATIONS FOR STUDIES OF CONSERVATION BIOLOGY

We have applied these spatial phylogenetic methods, while further developing them, to the floras of many parts of the world, including Australia (Thornhill et al. 2016, González-Orozco et al. 2016), Chile (Scherson et al. 2017), Florida (Allen et al., 2019), Norway (Mienna et al. 2019), and all of North America (Mishler et al. 2020). We found that not only do they help with understanding academic issues of ecology and evolution, they lend themselves to applied studies of conservation as well.

Understanding patterns of biodiversity is critical for conservation planning, given the urgent need to prioritize efforts in the face of rapid habitat loss and human-induced climate change. Biodiversity is almost always measured by counting species within a region to identify areas of particularly high species diversity as targets for conservation. So… if we are going to eliminate the species rank, then how are we to proceed in conservation? By this point in the book you can guess the answer: make use of the phylogeny. Biodiversity is not just the named species – it is the whole tree of life. So instead of counting species to measure biodiversity, let's use the tree itself as a measuring device via the methods of spatial phylogenetics. Rarity traditionally has meant not having many living close relatives. We can now quantitatively define what we mean by "many" and "close."

Since we have done comprehensive studies of the California flora, I will use this as my primary example. Baldwin et al. (2017) looked carefully at the issue of justifying the grid cell size and at potential biases in herbarium data, while presenting patterns of species richness and endemism, This non-phylogenetic study was done as a baseline to allow contrasts with a second study using the same spatial data but employing spatial phylogenetics (Thornhill et al. 2017). The latter analyses showed that examining the geographic distributions of branch lengths in a statistical framework did indeed add a new dimension which, in comparison with climatic data, helps to illuminate causes of endemism. In particular, the concentration of significant PE seen in more arid regions of California extends previous ideas about aridity as an evolutionary stimulus. The patterns seen were largely robust to phylogenetic uncertainty and time calibration but were sensitive to the use of occurrence data versus modeled ranges, indicating that special attention toward improving geographic distributional data should be a top priority in the future for advancing understanding of spatial patterns of biodiversity.

Kling et al. (2018) then conducted a thorough conservation assessment for the flora of California, developing and applying a novel phylogenetic algorithm that implements the principle of *complementarity*. This is the principle that one should take into account what has already been protected, when choosing the next location to protect. The priority in choosing the next location is to pick the place that maximizes the total biodiversity protected, thus the top choice will tend to be quite different from what has been protected before. Using this algorithm, grid cells are selected in order of conservation priority based on having poor protection, high intactness of natural vegetation, and high biodiversity value (i.e., many resident taxa with long phylogenetic branches that have small ranges and poor protection across their ranges). This study provided the most sophisticated "gap analysis" currently available, and we have been working since with the California Native Plant Society, the Nature Conservancy, and others to inform conservation efforts in the state. This approach implements an objective, rankless approach to conservation, and goes far beyond what could be done with the distribution of taxa alone, by using a phylogeny as its main measuring device.

IMPLICATIONS FOR NATURAL HISTORY

I described above (in Chapter 4) how formal taxonomic work would be done using a rankless approach at the former species level. But those who work on the practical side of systematics may be nervous at moving to rankless classification at this level. How can we teach people about nature without species? How can we do biodiversity inventories, record occurrence observations, write field guides, etc.?

Let's discuss teaching first. I know from personal experience while leading nature walks and teaching field courses that the first thing students and members of the public want to know is the name of a species they encounter. What is that bird? What is that tree? The public has a strongly typological view of species, many from a religious perspective ("species are the things God made"), some from a simple scientific perspective ("species are the basic kinds of life, separated by reproductive barriers"), and almost none with a more sophisticated scientific understanding of the actual evolutionary complexity of lineages discussed in this book.

Does the fact that people, from all cultures ranging from modern urbanites to traditional hunter-gatherers, strongly believe in the existence of species (Berlin 1973, Atran 1999) mean that scientists need to believe in species also? Shockingly, I have often heard that argument made by scientists, including some of my personal heroes (e.g., Gould's 1980 "a quahog is a quahog" argument). But really folks, public belief is a very poor guide to scientific truth, and scientists should know this. People believe in astrology; they don't believe in climate change. Should astronomers and ecologists throw up their hands and go with the public consensus? Of course not, and neither should evolutionary biologists. Our first responsibility as scientific educators should be to convey what we think is true about the world, and the methods we use to decide (for the moment) that something is true. We should not cater to ignorance, even if it is easier to do so.

Students and the public can handle the truth about species if it is explained to them. I find, for example, that it is easy to explain rankless classification in a public workshop or a biology class, which I do first, before getting into species. Anyone who has worked with classification at all knows how arbitrary it is to say something is a genus instead of a subgenus or a class instead of an order. After I have convinced them of this principle in general, then I introduce the topic of species – what about the species rank? Isn't it arbitrary also? I find that understanding comes easily, without much prompting.

Teaching biodiversity using the tree of life as a guide, rather than presenting a long, hierarchically ranked list of taxa to memorize, is not only more accurate scientifically, it is more interesting and easier to assimilate. The former approach gives people a visual framework (like a roadmap of a city) to organize their knowledge about characteristics of organisms; the latter approach is akin to asking someone to memorize a phone book listing addresses of people in that city. Teaching the fluidity of processes responsible for the clades that we observe in nature is a good way to get people to understand and accept evolution rather than keep thinking like a creationist.

What about printed natural history guides, websites, or social media apps that are meant to help people learn about biodiversity, identify taxa they encounter, and record their observations? Do you need ranked classifications including species for these purposes? We do need names for groups for sure, but we certainly don't need the ranks; e.g. it doesn't matter if a distinctive flowering plant is considered a species or a subspecies, if it can be told apart from its relatives.

In many taxonomic groups, it is very difficult to identify a specimen down to fine taxonomic scales without a microscope and knowledge of technical terms, or a DNA sequence, so it is common under the current ranked codes of nomenclature to have a field guide "stop" at an entry for a higher taxon. For example, Powell & Hogue's (1979) *California Insects* goes down to the species level in many butterflies because they are well-known and easy to tell apart, but in many other groups, the "terminal" treatments are for higher-level taxa, e.g., Springtails (order) or Fruitflies (family). There is absolutely nothing wrong with this; it is a rational adjustment to the capabilities and goals of the users of the guide. But note that "Springtail" (or "Collembola") works fine as a name for identification; the fact that it is ranked as an order is irrelevant to whether you can recognize it.

The best way to think about the act of identifying an organism is as the act of placing it on a phylogenetic tree. It does not need to be attached to a terminal branch for a person to gain some inferences about its traits and biology. Knowing something is an orchid, whale, liverwort, spider, or a springtail is a start for a beginner, and helps them begin to understand nature even in a new part of the world they might travel to. As a beginner learns and becomes more advanced in knowledge about a group, they can handle distinguishing finer clades (e.g., that liverwort is a *Riccia*!). No ranks are needed in the process of learning about nature, just a mental model of the tree of life that one keeps enriching by learning new branches.

SUMMARY

In all these areas of application, ranging from academic science to practical uses of classification in everyday life, matching the naming system as closely as possible to the evolutionary processes driving observed biodiversity patterns is beneficial. We get a more useful classification by incorporating our best models of biological reality (i.e., phylogenies) in them. Natural evolutionary processes have no "ranks" at any level – including the one formerly known as species – lineages diverge from each other (and sometimes reticulate again) at many nested levels, all of which are interesting from one standpoint or another. We should name the sufficiently diverged lineages with a rankless uninomial and can keep track of the nesting of such clade-based names mentally (to the extent that we personally "know" a group) and comprehensively in databases used to enable comparative evolutionary studies, ecological applications, biogeographic algorithms, conservation assessment, identification guides, and teaching materials. The species rank no longer makes sense in the biological world as we know it today; it needs to disappear along with all other taxonomic ranks.

LITERATURE CITED

Allen, J.M., C.C. Germain-Aubrey, N. Barve, K.M. Neubig, L.C. Majure, S.W. Laffan, B.D. Mishler, H.L. Owens, S.A. Smith, W.M. Whitten, J.R. Abbott, D.E. Soltis, R. Guralnick, and P.S. Soltis. 2019. Spatial phylogenetics of Florida vascular plants: the effects of calibration and uncertainty on diversity estimates. *iScience* 11: 57–70.

Atran, S. 1999. The universal primacy of generic species in folk biological taxonomy: Implications for human biological, cultural and scientific evolution. Pp. 231–261 in: *Species, New Interdisciplinary Essays*, R. A. Wilson (ed.). Bradford/MIT Press, Cambridge, MA.

Avise, J.C. 1989. Gene trees and organismal histories: a phylogenetic approach to population biology. *Evolution* 43: 1192–1208.

Baldwin, B.G., A.H. Thornhill, W.A. Freyman, D.D. Ackerly, M.M. Kling, N. Morueta-Holme, and B.D. Mishler. 2017. Species richness and endemism in the native flora of California. *American Journal of Botany* 104: 487–501.

Berlin, B. 1973. Folk systematics in relation to biological classification and nomenclature. *Annual Review of Ecology and Systematics* 4: 259–271.

Doyle, J.J., J.L. Doyle, J.T. Rauscher and A.H.D. Brown. 2004. Diploid and polyploid reticulate evolution throughout the history of the perennial soybeans (*Glycine* subgenus *Glycine*). *New Phytologist* 161: 121–132.

Faith, D.P. 1992. Conservation evaluation and phylogenetic diversity. *Biological Conservation* 61: 1–10.

González-Orozco, C.E., L.J. Pollock, A.H. Thornhill, B.D. Mishler, N. Knerr, S.W. Laffan, J.T. Miller, D.F. Rosauer, D.P. Faith D.A. Nipperess, H.Kujala, S. Linke, N. Butt, C. Külheim, M.D. Crisp, and B. Gruber. 2016. Phylogenetic approaches reveal biodiversity threats under climate change. *Nature Climate Change* 6: 1110–1114.

Gould, S.J. 1980. A quahog is a quahog. Pp. 204–213 in: *The Pandas Thumb*. Norton.

Gould, S.J., N.L. Gilinski, R.Z. German, 1987. Asymmetry of lineages and the direction of evolutionary time. *Science* 236: 1437–1441.

Gustafson, D.J., D.J. Gibson, and D.L Nickrent. 2014. Using local seeds in prairie restoration—data support the paradigm. *Native Plants Journal* 6(1): 25–28.

Grant, V. 1981. *Plant Speciation*. Columbia University Press, New York.

Hutchinson, G.E. 1965. *The Ecological Theater and The Evolutionary Play*. Yale University Press, New Haven.

Kling, M.M., B.D. Mishler, A.H. Thornhill, B.G. Baldwin, and D.D. Ackerly. 2018. Facets of phylodiversity: evolutionary diversification, divergence, and survival as conservation targets. *Philosophical Transactions Royal Society B*. 374: 20170397.

Laffan, S.W., D.F. Rosauer, G. Di Virgilio, J.T. Miller, C.E. González-Orozco, N. Knerr, A.H. Thornhill, and B.D. Mishler. 2016. Range-weighted metrics of species and phylogenetic turnover can better resolve biogeographic transition zones. *Methods in Ecology and Evolution* 7: 580–588.

Mayr, E. 1982. Speciation and macroevolution. *Evolution* 36: 1119–1132.

McKay, J.K., C.E. Christian, S. Harrison, and K.J Rice. 2005. How local is local?—a review of practical and conceptual issues in the genetics of restoration. *Restoration Ecology* 13: 432–440.

Mienna, I.M., J.D.M. Speed, M. Bendiksby, A.H. Thornhill, B.D. Mishler, and M.D. Martin. 2019. Differential patterns of floristic phylogenetic diversity across a post-glacial landscape. *Journal of Biogeograph* 47: 915–926.

Miller, J.T., G. Jolley-Rogers, B.D. Mishler, and A.H. Thornhill. 2018. Phylogenetic diversity is a better measure of biodiversity than taxon counting. *Journal of Systematics and Evolution* 56: 663–667.

Mishler, B.D. 2010. Species are not uniquely real biological entities. Pp. 110–122 in: *Contemporary Debates in Philosophy of Biology*, F. Ayala and R. Arp (eds.). Wiley-Blackwell, Weinheim, Germany.

Mishler, B.D., N.J. Knerr, C.E. González-Orozco, A.H. Thornhill, S.W. Laffan, and J.T. Miller. 2014. Phylogenetic measures of biodiversity and neo- and paleo-endemism in Australian *Acacia*. *Nature Communications* 5: 4473.

Mishler, B.D., R.Guralnick, P.S. Soltis, S.A. Smith, D.E. Soltis, N. Barve, J.M. Allen, and S.W. Laffan. 2020. Spatial phylogenetics of the North American flora. *Journal of Systematics and Evolution* 58: 393–405.

Nixon, K.C. and Q.D. Wheeler. 1990. An amplification of the phylogenetic species concept. *Cladistics* 6:211–223.

Nürk, N.M., H.P. Linder, R.E. Onstein, M.J. Larcombe, C.E. Hughes, L.P Fernández, P.M. Schlüter, L. Valente, C. Beierkuhnlein, V. Cutts, M.J. Donoghue, E.J. Edwards, R. Field S.G.A. Flantua, S.I. Higgins, A. Jentsch, S. Liede-Schumann, and M.D. Pirie. 2020. Diversification in evolutionary arenas—assessment and synthesis. *Ecology and Evolution* 10: 6163–6182.

Powell, J.A. and C.L. Hogue. 1979. *California Insects*. University of California Press, Berkeley.

Rosauer D, S.W. Laffan, M.D. Crisp, S.C. Donnellan, and L.G. Cook. 2009. Phylogenetic endemism: a new approach for identifying geographical concentrations of evolutionary history. *Molecular Ecology* 18: 4061–4072.

Scherson, R.A., A.H. Thornhill, R. Urbina-Casanova, WA. Freyman, P.A. Pliscoff, and B.D. Mishler. 2017. Spatial phylogenetics of the vascular flora of Chile. *Molecular Phylogenetics and Evolution* 112: 88–95.

Thornhill, A.H., B.D. Mishler, N. Knerr, C.E. Gonzalez-Orozco, C.M. Costion, D.M. Crayn, S.W. Laffan, and J.T. Miller. 2016. Continental-scale spatial phylogenetics of Australian angiosperms provides insights into ecology, evolution and conservation. *Journal of Biogeography* 43: 2085–2098.

Thornhill, A.H., B.G. Baldwin, W.A. Freyman, S. Nosratinia, M.M. Kling, N. Morueta-Holme, T.P Madsen, D.D. Ackerly, and B.D. Mishler. 2017. Spatial phylogenetics of the native California flora. *BMC Biology* 15:96.

Webb, C.O., D.D. Ackerly, M.A. McPeek and M.J. Donoghue 2002. Phylogenies and community ecology. *Annual Review of Ecology and Systematics* 33: 475–505.

Endnotes

1. B.D. Mishler and M.J. Donoghue. 1982. Species concepts: a case for pluralism. Systematic Zoology 31: 491–503. [reprinted by permission] (P.11)

2. Gould (1979) and others have defended the biological species concept on the grounds that the same taxa recognized by western taxonomists are recognized by tribespeople in New Guinea, etc. There are several problems with this kind of argument. First, it is not clear that this finding constitutes an independent test because, after all, New Guinea tribespeople are human too, with similar cognitive principles and limitations of language. It should also be borne in mind that the observer is by no means neutral. Folk taxonomies have been collected by people with a knowledge of evolution and modern systematic concepts. Second, it is generally not a strong argument to show that a pre-scientific society has recognized something that modern science currently accepts. Surely a modem astronomer would not consider it very strong evidence that a primitive mythology supported one cosmological theory over another. Finally, the taxa recognized by western taxonomists (and often by natives at some level of their linguistic hierarchy) in these instances are not known to be biological species – for the most part, they are morphological units that are believed to be reproductively isolated from other such units. (P.13)

3. Initially, the biological species concept was embraced and promulgated by plant systematists interested in evolution (Stebbins, 1950; Grant, 1957). Cronquist (1978) detailed Grant's efforts (from 1956 to 1966) to apply the biological species concept in *Gilia* (Polemoniaceae). It very soon became apparent that the biological species concept was fraught with difficulties, but Grant chose to amend the concept (rather than abandon it altogether), first (1957) with the notion of the syngameon (i.e., the unit of interbreeding higher than the species), later (1971) by adopting an evolutionary species concept. Finally, in the second edition of his classic book on plant speciation, Grant (1981) treats species in a more flexible and pluralistic manner. Some botanists (e.g., Stebbins, 1979:25) continue to feel that the biological species concept, or some modification of it, is the only suitable framework for understanding plant diversity. However, many (perhaps most) botanical systematists remain rather skeptical about the general applicability of the concept in botany (Davis and Heywood, 1963; Raven, 1976; Cronquist, 1978; Levin, 1979; Stevens, 1980a).

    The different attitudes of zoologists and botanists towards the concept of species may be of interest to historians, sociologists, and philosophers of science. For organismic and evolutionary biology the "modern synthesis" of the 1930's and 1940's may have represented a revolution in the sense of Kuhn (1970). For systematists, the principal outcome was the biological species concept. Zoologists (especially vertebrate systematists) appear to have largely accepted the new paradigm and to have entered a period of "normal science' applying the concept in particular cases ("puzzle-solving"). While problems like sibling species, semispecies, and subspecies have become apparent, these have generally not prompted a critical evaluation of the paradigm or a proliferation of alternatives. In contrast, in the botanical community the biological species concept was soon found to be inapplicable or of difficult application and likely to lead to confusion. This resulted in a groping for alternatives and a defense of older concepts. In this regard, the historical development of species concepts in botany seems to fit better Feyerabend's (1970) characterization of scientific change as the simultaneous practice of normal science and the proliferation of alternative theories. (P.13)

4. The zoologists initially responsible for developing the biological species concept were aware of the difficulties in applying the concept in some groups of animals and many

groups of plants. Dobzhansky (1937, 1972) consistently pointed out the diversity of "species situations" observable in nature. Mayr (1942:122) was careful to point out differences between plants and animals, and difficulties in the practical application of the biological species concept in some cases. Particularly rigid versions of the biological species concept have been promulgated more recently, in attempted generalizations that have shown a startling lack of concern for the biology of the majority of organisms on earth. Mayr (1982) has examined the resistance of botanists to the biological species concept and concluded that "the concept does not describe an exceptional situation" (p. 280). But he grants some justification to the ideas of "certain botanists" who question "whether the wide spectrum of breeding systems that can be found in plants can all be subsumed under the single concept (and term) 'species'" (p. 278). (P.14)

5. The species concepts of Cronquist and of Nelson and Platnick are as follows:
   Cronquist (1978:15): "the smallest groups that are consistently and persistently distinct, and distinguishable by ordinary means."

   Nelson and Platnick (1981:12): "the smallest detected samples of self-perpetuating organisms that have unique sets of characters." (P.20)

6. Initially the "gene" was considered to be the unit of heredity, but the classical concept of gene has been replaced by several concepts which stand in a complex relation to one another (Hull, 1965). The use of a disjunctive definition (Hull, 1965) allows a single term to designate a complex of concepts. However, this can become so confusing that it may be desirable to replace (at least in part) an old terminology with a new set of terms with more precise meanings. (P.21)

7. B.D. Mishler. 1985. The morphological, developmental, and phylogenetic basis of species concepts in bryophytes. The Bryologist 88: 207–214. [reprinted by permission] (P.26)

8. B.D. Mishler and A.F. Budd. 1990. Species and evolution in clonal organisms – introduction. Systematic Botany 15: 79–85. [reprinted by permission] (P.39)

9. B.D. Mishler and R.N. Brandon. 1987. Individuality, pluralism, and the phylogenetic species concept. Biology and Philosophy 2: 397–414. [reprinted by permission] (P.54)

10. We should note at the outset that, contrary to the impression one is likely to get from the literature on species-as-individuals, the class-individual distinction is not a distinction taken directly from logic. First, Hull and Ghiselin are using a restricted notion of classes. Something counts as a class for them only if its membership can be specified in a spatiotemporally unrestricted way. Logic places no such restriction on classes. Although Hull (1978) is reasonably clear on this point, not everyone else has been and this has lead to some confusion. Second, the operative notion of "individual" comes more from common sense zoology than from logic. (P.55)

11. As pointed out by Hull (pers. comm.), when the distinction between grouping and ranking has previously been made, it was often blurred. This may often be because researchers use variations on the same theme for both grouping and ranking; e.g., patterns of morphological similarity or of gene exchange. As will be apparent below, we advocate distinctly different criteria for grouping than for ranking. (P.60)

12. A similar result has been arrived at by Holman (pers. comm.) based on comparisons between bdelloid rotifers (which are exclusively parthenogenic) and monogonont rotifers (which occasionally reproduce sexually). Using numbers of synonymous species names as an index of taxonomic distinctness of species, he has shown that bdelloid species are apparently more consistently recognized by taxonomists than are monogonont species. (P.63)

13. B.D. Mishler and E. Theriot. 2000. The phylogenetic species concept sensu Mishler and Theriot: monophyly, apomorphy, and phylogenetic species concepts. In Q.D.

Wheeler & R. Meier (eds.), Species Concepts and Phylogenetic Theory: A Debate, pp.44–54. Columbia University Press. [reprinted by permission] (P.70)

14. B.D. Mishler. 1999. Getting rid of species? In R. Wilson (ed.), Species: New Interdisciplinary Essays, pp.307–315. MIT Press. [reprinted by permission] (P.89)

15. In Hennigian phylogenetic systematics, "homology" is defined historically as a feature shared by two organisms because of descent from a common ancestor that had that feature. (P.92)

16. A strictly monophyletic group is one that contains all and only descendants of a common ancestor. A paraphyletic group is one the excludes some of the descendants of the common ancestor. (P.92)

17. Note that some of the nested clades will have a formal suffix indicating their previous rank (e.g., "-idae" for family). While these ending would be retained for exiting clade names, in order to avoid confusion, there would be no meaning attached to them and newly proposed clade names would have no particular suffix requirement. (P.94)

18. B.D. Mishler. 2010. Species are not uniquely real biological entities. In F. Ayala and R. Arp (eds.), Contemporary Debates in Philosophy of Biology, pp. 110–122. Wiley-Blackwell. [reprinted by permission] (P.97)

19. N. Cellinese, D.A. Baum, and B.D. Mishler. 2012. Species and phylogenetic nomenclature. Systematic Biology 61: 885–891. [reprinted by permission] (P.111)

20. B.D. Mishler and J.S. Wilkins. 2018. The hunting of the SNaRC: a snarky solution to the species problem. Philosophy, Theory, and Practice in Biology. 10: 1–18. [reprinted by permission] (P.124)

21. The observation is due to Jody Hey (pers. comm.). Contrary to Mayr's and others' characterizations, Darwin did not intend to define species, but to explain why they existed. (P.126)

22. It is misleadingly also known as the "Unified Species Concept." This term is misleading because, although all species to the extent they are natural objects form lineages, that is also true of taxa at all ranks. It is entirely unclear what kind of lineages uniquely qualify to be named at the species rank. So this conception is unified (and general) just to the extent that it proposes a necessary but insufficient criterion for a natural species concept. (P.127)

23. "Good species" form the proof of concept for biologists for concepts of species (the rank). Every biologist knows what form good species take in their specialty, but each subdiscipline differs in subtle or gross ways from other subdisciplines. See Amitani (2015) for a discussion of this and a characterization of "good species" as a form of prototypical reasoning. (P.127)

24. Numerical taxonomy, also known as the phenetics school (from the Greek phaineros for "appearance") classified groups according to their "overall similarity." This fell prey to the problems discussed by Nelson Goodman; as he says, similarity is cheap (see Decock and Douven [2011] for a discussion): "Similarity, I submit, is insidious. And if the association here with invidious comparison is itself invidious, so much the better. Similarity, ever ready to solve philosophical problems and overcome obstacles, is a pretender, an impostor, a quack. It has, indeed, its place and its uses, but is more often found where it does not belong, professing powers it does not possess." (Goodman 1972, 437) Depending on the characters used, phenetic groups, known as Operational Taxonomic Units or OTUs, could contradict other analyses using different characters of the same organisms. (P.128)

25. Cladism is the approach to classification that defines taxa by uniquely shared common ancestry (monophyly), as evidenced by shared derived characters. It is also known as phylogenetic systematics. (P.128)

26. Existing species concepts (except the conventional ones) define species in terms of some model or process, which is to say, as entities of a particular theoretical kind. To treat species as phenomena in need of explanation is to not beg the question in favor of a prior mechanism, which we take to be a scientific virtue. Thanks to a reviewer for raising this question. (P.128)

27. Which traits are selected to use for such comparisons depend a lot on prior experience rather than theoretic criteria, in traditional societies as well as in modern taxonomy. (P.129)

28. See Scerri (2007) for an example from chemistry, the periodic table. The properties of elements were experimentally measured and the periodicity of these properties noted before any theoretical explanation (such as valency theory or electron shells and proton number) was available. Likewise, plate tectonics was observed as a phenomenal pattern before an explanation was offered (Oreskes and LeGrand 2003). (P.129)

29. There is an extensive literature on folk taxonomy. We simplify here and are not suggesting that the same basic taxa are recognized in all or even most cultures (Berlin 1973, 1976; Berlin, Breedlove, et al. 1973; Durkheim and Mauss 1963; Medin and Atran 1999; Sousa, Atran, et al. 2002; Zachar 2000). (P.129)

30. In taxonomy, a natural group is a real fact about the organisms' interrelationships. It is contrasted to artificial groups, which are merely conventional, or based upon arbitrary characters. (P.130)

31. http://www.phyloregnum.org/ (P.130)

32. If monophyly is the key, then the name or structure of the including clade is not relevant to the identification of that smallest (currently known) clade. Hence a genus name, giving a binomial, is not necessary. (P.131)

33. This is similar in some respects to Pleijel and Rouse's notion of a Least Inclusive Taxonomic Unit, or LITU (Pleijel and Rouse 2000), in that there is no rank that is fundamental. They say "Identification of taxa as LITUs are statements about the current state of knowledge (or lack thereof) without implying that they have no internal nested structure." However, with SNaRCs we restrict the naming of terminal taxa to entities regarded as clades, by requiring the author of a SNaRC name to present evidence of monophyly. (P.131)

34. Horizontal genetic transfer can occur in a number of ways, summarized by (Soucy, Huang, et al. 2015). However, it is notable that this usually offers a different phylogeny and recovers different taxa for the gene tree, not the taxon tree (Degnan and Rosenberg 2009). (P.132)

35. This is to be expected since monophyly is only a relatively recent desideratum (for some) in taxonomy, many species were named before the cladistics revolution, and a large number of practicing taxonomists still do not accept monophyly as a desideratum. (P.133)

36. A recent paper now assigns specific status, *Pongo tapanuliensis*, to a central Bornean population (Nater, Mattle-Greminger, et al. 2017). (P.134)

Index